Emerald Architecture

CASE STUDIES IN GREEN BUILDING

MCGRAW-HILL'S GREENSOURCE SERIES

Gevorkian
Solar Power in Building Design: The Engineer's Complete Design Resource

GreenSource: The Magazine of Sustainable Design
Emerald Architecture: Case Studies in Green Building

Haselbach
The Engineering Guide to LEED-New Construction: Sustainable Construction for Engineers

Emerald Architecture

CASE STUDIES IN GREEN BUILDING

GREENSOURCE: THE MAGAZINE OF SUSTAINABLE DESIGN

NEW YORK   CHICAGO   SAN FRANCISCO   LISBON   LONDON   MADRID
MEXICO CITY   MILAN   NEW DELHI   SAN JUAN   SEOUL
SINGAPORE   SYDNEY   TORONTO

Emerald
Architecture
CASE STUDIES IN GREEN BUILDING

1 2 3 4 5 6 7 8 9 0   WCK/WCK   0 1 4 3 2 1 0 9 8

ISBN 978-0-07-154411-5
MHID 0-07-154411-9

Sponsoring Editor: Joy Bramble Oehlkers
Editing Supervisor: Stephen M. Smith
Production Supervisor: Richard C. Ruzycka
Editors: Charles Linn, FAIA; Jane Kolleeny; Russell Fortmeyer; Joann Gonchar, AIA
Design and Composition: Mitch Shostak, Corey Kuepfer / Shostak Studios; Clifford Rumpf, Senior Designer
Art Director, Cover: Jeff Weeks

Printed and bound by Quebecor/Versailles.

McGraw-Hill books are available at special quantity discounts to use as premiums and sales promotions, or for use in corporate training programs. To contact a special sales representative, please visit the Contact Us page at www.mhprofessional.com.

♲ This book is printed on recycled, acid-free paper made from 100% postconsumer waste.

Library of Congress Cataloging-in-Publication Data

Emerald architecture : case studies in green building / GreenSource,
    the magazine of sustainable design.
            p.   cm.
     Includes index.
     ISBN 978-0-07-154411-5 (alk. paper)
      1. Sustainable buildings—Design and construction—Case studies.
   2. Sustainable architecture—Case studies.   3. Sustainable buildings—
   Pictorial works.   I. GreenSource.
   TH880.E47   2008
   720'.47—dc22   2008005557

Contents

FOREWORD

O NE CANNOT BE AN ARCHITECT WITHOUT UNDERSTANDING THAT THE IMPACT OF buildings on the environment is profound, and certainly awareness of this fact has only deepened over the course of my 40-year career. At the outset of my journey, my peers and I thought mainly about how our buildings affected their sites and communities. We were not yet able to appreciate the consequences that the millions of buildings we were building in our country had on a global scale. Issues like resource depletion, or the effects of pollution that occurred as a consequence of our work, would come to our attention from time to time, but generally we discounted any negative outcomes or accepted them as the cost of doing business.

As it happens, our profession is happiest and healthiest when we are doing what we do very well: creating the built environment our population and its economy needs to be happy, healthy, and prosperous. Since World War II, despite a few periods of economic decline and uncertainty, American society has enjoyed a period of growth and wealth unparalleled in human history. At the same time, however, our building and community designs have set global benchmarks for consumption, waste, and pollution. Indeed, scientists have issued a series of urgent reports putting us on notice that global climate change threatens human life on earth, and have identified buildings as the single largest contributor to climate change.

One of the advantages of our ever-more-connected world is that we are better able to share information, and understand what scientists are telling us about the damage our buildings do to the environment. We now have the technology to assess the effects and effectiveness of what we do, which may help to convince the skeptics that we are making a difference.

Because buildings and the influences that indirectly radiate from them have such a vast impact on our future, it is a certainty that architects can make a real difference. I am proud to say that I believe that architects and engineers tend to be an inherently honest and ethical group of people. I have never met any who would intentionally inflict harm on our environment. However, one problem that we have always had is that designing and making buildings and communities is a complex enterprise. The changes in building technology alone can be challenging but the human and environmental issues are even more daunting. Can we develop new design approaches to manage these new challenges and complexities to reveal exactly how to do the right thing?

We are now well on our way. The establishment of the U.S. Green Building Council (USGBC) and the creation of the Leadership in Energy and Environmental Design (LEED) method of verifying that new buildings are green have been transformative events. These benchmarks have led the way toward a new kind of standard practice in the marketplace, showing the design-build community what it should be doing, as well as measuring in very real terms the benefits of its efforts.

The rating system has been extremely helpful in enlightening those in the building profession and construction industry, not to mention the public, that they can easily make a difference, often without spending any extra money. The USGBC's idea was that by establishing metrics it could transform the market for green architecture from something that was almost unknown to being standard practice.

But without support, market transformations can still fail, no matter how worthy. Architects and engineers have had to learn about sustainable design quickly, and they do not work alone. To do green buildings, they frequently have to excite and inspire their clients and other stakeholders. *GreenSource* magazine has played an indispensable part in making the point that environmentally responsible architecture is necessary and achievable. Its case studies do this not simply by repeating the claims touted in press releases issued by owners and developers, but by showing real data that has been collected by architects and engineers during the process of building their buildings. This book, *Emerald Architecture*, is a collection of 24 of the best case studies from *GreenSource's* first two years of publication. It is difficult to overestimate the amount of influence the magazine has had on the profession's steady embrace of green principles. Whether you are an architect or engineer looking for design methodology, a contractor who wants to learn from other builders, or an owner who wants to understand whether sustainable architecture can be applied to your project, you will find answers here.

We have only begun to tap sustainable architecture's potential for strengthening our social, economic, and environmental vitality. To this end, *Emerald Architecture* is a tremendous resource.

—ROBERT BERKEBILE, FAIA

Robert Berkebile, FAIA, is a leader of an international effort by American architects to develop the information the design and construction industry needs to create sustainable communities. He is the founding chairman of the American Institute of Architects Committee on the Environment and has served on the board of the U.S. Green Building Council. He is a principal of BNIM Architects in Kansas City, Missouri.

INTRODUCTION

THE CASE STUDIES THAT COMPRISE MOST OF *EMERALD ARCHITECTURE'S* PAGES COME from *GreenSource: The Magazine of Sustainable Design*, launched by McGraw-Hill Construction and BuildingGreen, Inc., in 2006. *GreenSource* provides architects, engineers, contractors, and building owners with the most accurate, comprehensive guidance available today for the design and construction of environmentally responsible buildings. Given the countless books and articles on sustainable design that have been published over the years, we determined our readers could best learn about what makes a green building successful from case studies of recently completed projects. These are the core of each issue of the magazine. In order to produce *Emerald Architecture*, we collected and updated the case studies, and organized them into chapters according to building type.

Every case study starts by explaining the needs and desires of the clients and building end-users. We have added weather charts, drawings, and illustrations of green aspects of the designs to the background information, and as well as lists of key project data, team members, and green products to supplement the narratives. Where possible, the case studies even include details of what did not work.

How do we know these buildings are green? The majority of buildings included in *Emerald Architecture* have been rated under the U.S. Green Building Council's LEED Rating System. In order to receive a LEED rating, designers and consultants must compile extensive documentation to verify the sustainable features of each building. As part of the certification process, project teams must provide predicted annual energy-performance data to the council for review. We have included this information or, when available, the actual performance data with each case study. Although the LEED Rating System has its critics, we believe it is the most reliable way of verifying the sustainable characteristics of buildings. Designers who submitted projects that were not LEED rated were still required to submit building-performance data.

To our knowledge no book contains as many deeply researched, richly illustrated, well-documented case studies as this one. In fact, it would be difficult for a single author to produce such a book within the time span that this one has been put together. That's why it is so advantageous to produce a book from a magazine: In this case, reader demands for up-to-date information ensured that the buildings in this publication have been completed very recently. At the time of *Emerald Architecture's* first printing, it had been less than six months since some of the case studies had been published in *GreenSource*.

ACKNOWLEDGMENTS

WE WOULD LIKE TO THANK ALL OF THE MEMBERS OF THE *GREENSOURCE* team including: Robert Ivy, FAIA, editorial director; Nadav Malin, executive editor; Jane Kolleeny, managing editor; Charles Linn, FAIA, and Alex Wilson, consulting editors; and Joann Gonchar, AIA, Russell Fortmeyer, Mark Piepkorn, Tristan Korthals Altes, and Jessica Boehland, contributing editors. Special thanks go to *GreenSource's* publisher, Mark Kelly; McGraw-Hill Construction's former group design director Anna Schlesinger; Mitch Shostak and Corey Kuepfer of Shostak Studios, who design *GreenSource* as well as this book; Clifford Rumpf, *GreenSource's* senior designer; and I-Ni Chen, *GreenSource's* illustrator.

Contributing writers for the case studies include Nadav Malin, Russell Fortmeyer, Joann Gonchar, AIA, Tristan Korthals Altes, and Jessica Boehland. Interns Henry Ng and Rebecca Ward wrote some of the introductions to the chapters and performed other important tasks. Additional contributors include Blair Kamin, Kira Gould, David Sokol, Randy Gragg, and B.J. Novitsky. David Delp assisted with copy-editing and proofreading.

Charles Linn initially proposed *Emerald Architecture* to Steve Chapman of McGraw-Hill Professional and Jay McGraw of McGraw-Hill Construction, and they helped make it a reality. Cary Sullivan, formerly of McGraw-Hill Professional, was an early champion, and McGraw-Hill Professional senior editor Joy Bramble Oehlkers saw the project through its publication. Jane Kolleeny was also the project manager for *Emerald Architecture*. The U.S. Green Building Council provides *GreenSource* to its members, and without their support the magazine would not have achieved the success it has today.

McGraw-Hill Construction is the world's leading provider of information for the construction industry. This division of McGraw-Hill, Inc., publishes *Architectural Record*, *Engineering News-Record*, *GreenSource*, and 12 regional construction magazines. The construction group also includes the Sweets Network, Dodge, which publishes information on building projects, and McGraw-Hill Construction Analytics, which conducts building industry research. BuildingGreen, Inc., is an independent research and publishing company responsible for several highly regarded resources for sustainable building design. It publishes *Environmental Building News* and the *GreenSpec Directory*, and produces the BuildingGreen Suite of online tools.

Chapter 1

A GREEN BUILDING PRIMER

T HE BULK OF *EMERALD ARCHITECTURE* CONSISTS OF CASE STUDIES OF SUSTAINABLE buildings. In order for readers to make the most of them, this chapter will answer questions such as, what are sustainable buildings, and what makes them different from other buildings? What do we mean by sustainability? What is integrated design? How were the buildings for this book chosen? What are the U.S. Green Building Council (USGBC) and the LEED Green Building Rating System? What do statistics like "annual energy purchased" or "annual carbon foot-print" mean? Why publish another book about green buildings?

THE EMERGING MARKET FOR SUSTAINABLE BUILDINGS

Building design professionals have long cared for the natural environment, since the work they do exists within the interconnected realm of an eco-system. Understanding a site and what a building does to it has been a primary concern of architects, engineers, landscape architects, and contractors, as well as many other professionals, for decades. Ecological imperatives came to the fore for many people with the 1962 publication of Rachel Carson's *Silent Spring*, which persuasively illustrated how the actions of humans had transformed the natural world in ways they never anticipated. Carson demonstrated that the use of pesticides in agriculture was damaging the natural environment, our food supply, and the health of some wild animal populations. It's hard to believe, in 2008, that this was radical thinking at one time. The influence of Carson's book in green design circles cannot be overstated and was evident in later books, like Ian L. McHarg's *Design With Nature*, from 1967, which focused primarily on placing cities within the larger contexts of regional watersheds and geological development.

Many other influential books that followed continued to develop a larger understanding of the challenges designers and builders face in a world where resources are decreasing as energy demand grows. Today, the U.S. Department of Energy estimates that buildings use 70 percent of the country's electricity supply, and we know that something about the way we build our world has to change. We have come to associate these concerns with the general subject of "sustainability."

Although sustainability has many definitions, one that gets at the idea behind the way this word is used in *Emerald Architecture* was written at the World Commission on Environment and Development in 1987: Sustainability means "development that meets the needs of the present without compromising the ability of future generations to meet their own needs." Taking the idea a step further, sustainable buildings are those that have been designed deliberately to have a minimal impact on the environment. A partial list of things designers might consider when doing a sustainable building includes decreasing resource consumption during construction and operation, the source and composition of a building's materials, and siting a building to minimize environmental damage and soil erosion. When properly designed, sustainable buildings can produce many benefits: Operating costs are lower than those of similar buildings, and occupants often report that they are more comfortable and productive in them. Post-occupancy studies, which have been rare in the world of architecture, increasingly show these benefits to be measurably true.

It has surely never been a surprise to any architect or engineer that a building intended for a given site and program can be designed in such a way that its impact on the environment is either minimal or possibly quite severe. Because most building design professionals took up their careers with the hope they might improve the world around them, many naturally tend toward using practices that are sustainable. But designers are seldom able to design buildings to please themselves alone. They answer to clients who are cost-conscious, who are sometimes driven by ambitions that encourage excess, or who simply don't understand how sustainable design works and how it can benefit them and the occupants of their buildings. It is difficult for architects and engineers to pursue a sustainability agenda without client support, and for this reason buildings that achieved a high level of sustainability were, until recently, relatively rare. Another issue that has discouraged sustainable buildings has been the difficulty in determining what sustainability goals for a particular building were realistic and, once it was finished, to gather the resources needed to measure whether they were met. For example, would the gains created by using large areas of glass to allow the sun to warm a building during winter days create unacceptable heat losses at night?

Fortunately, designers are now able to use software that simulates building performance under many conditions to predict such effects and help them make informed judgments.

Attempting to assess the effect of a building at a global scale seems impossible. For example, is the harvesting of renewable timber better or worse for the environment than the manufacture of recyclable steel? Determining the answers to these questions is beyond the scope of this book; suffice it to say, being a competent designer of sustainable buildings requires an awareness of all these kinds of issues. Thankfully, more resources are available now to help inform designers and their clients today than there were at any previous time.

In recent years, several developments have driven the desire of a few members of the public that sustainable design should be the standard, rather than an exceptional way of working, a change architects could not have willed on their own. For one thing, there recently has been a general acceptance of the evidence that there is a connection between the greenhouse gases released during the production of energy and rapid, permanent increases in the Earth's temperature. Most people don't need a film like Al Gore's *An Inconvenient Truth* to understand that global warming is real—they have felt it for themselves as the 10 hottest years on record for the U.S. have occurred since 1990. As well, the larger public has become aware that buildings play an important part in contributing to these changes. Their operation requires energy, which in turn creates greenhouse gases. That has been accompanied by a greater understanding of how the built environment affects the greater world in a holistic sense: Buildings consume water and expel sewage, and their construction claims irreplaceable habitat, as can the mining of minerals and harvesting of timber used to manufacture building materials. Building-materials manufacturing and transport also require that vast amounts of fossil fuels be burned.

These problems aren't insurmountable, but they are substantial. They require a change in the way those in the building-industry professions do their business. Change will require the education and re-education of great numbers of architects, engineers, consultants, and contractors, in addition to manufacturers, suppliers, scientists, maintenance staff, and building occupants—basically, everyone on the planet has to sit up and listen. It's certainly a dynamic time to be in the business of making buildings, but only if one recognizes the challenge that no longer lies ahead but right under our feet.

INTEGRATED DESIGN

Since the 1940s and 1950s, as new technology is introduced into our buildings to help make them perform better and more efficiently, specialist consultants are emerging in a range of fields related to design and construction: telecommunications consultants, façade consultants, acousticians, lighting designers, energy modelers, fire engineers, vertical transportation consultants, and environmental consultants, to name a few. Specialists' knowledge in these fields has helped to transform how buildings engage the environment, providing many solutions—like daylight dimming systems—we take for granted in high-performance green buildings. But in order to realize the apparently limitless range of possibilities for high-performance buildings, the design industry has embraced a concept of collaboration that has come to be known as "integrated design."

Integrated design places each of these consultants at the design table with architects and engineers from a project's inception. Whereas previous design models had the architects develop a scheme that was then passed onto building services consultants—like structural, mechanical, electrical, and plumbing (SMEP) engineers—for design and coordination, nowadays the architects are involving people like the mechanical engineer in decisions that determine such things as what shape the building will have, whether the curtain wall is fitted with tinted or clear glass, whether internal or external shades will be installed, and even whether the building will have air-conditioning or will rely on operable windows for natural ventilation. These sorts of decisions, which influence the performance of green buildings, need the focused attention and open process that integrated design allows.

Project Checklist

This is the checklist that would be used for a LEED for New Construction project. The "v2.2" designation indicates that the building is going for a rating under version 2.2 of the LEED Rating System, which has been refined in order to make use of criticisms and comments made by users in the years since it went into use. The list is an excellent outline of all the elements the designers of a sustainable building should consider even if they are not attempting to receive a LEED rating for it.

Many, if not all, of the projects in *Emerald Architecture* are products of an integrated design approach. For example, the Des Moines Central Library, designed by David Chipperfield Architects, included a team of London architects and engineers collaborating with local architects and engineers to design and construct a copper-infused glass curtainwall that defines the building's exterior. The curtainwall, developed with the active participation of its manufacturer, consists of triple-glazed, 4-by-14-foot glass panels that encase fine, expanded copper mesh. The implications of this system rippled through nearly every other decision in the building's design, determining the type of the structural system used, the massing of the building, the mechanical system, and lighting. This one decision was made after weeks of analysis and investigation on the part of the team, leading to what may be characterized as "whole-systems thinking," where interactions among building elements are taken into account in order to exploit their synergies. Integrated design favors this front-loaded design process, where the implications of choosing building systems and materials are thought through before final decisions are made, giving designers and consultants time to explore the range of possibilities afforded by several options without locking themselves into one design approach.

Conventional wisdom has it that few projects can absorb this investment in design energy, particularly for schemes that may not work out in the end due to budget issues. Naysayers also point out that integrated design requires more face-to-face meetings for the design team at the beginning of a project, which can force up costs. But many sustainable design consultants swear otherwise. They say research and analysis at the beginning of a project can often save money in construction since it most likely requires fewer meetings and less problematic coordination throughout the life of the project. They argue that it is more an issue of where the project's budget is allocated than whether there is enough money for green design. Of course, like the design of any building, there are exceptions to this argument.

Another approach of integrated design is in considering the "end-use, least-cost" of a system when plan-

LEED for New Construction v2.2
Registered Project Checklist

Project Name:
Project Address:

	Yes	?	No				Points
					Sustainable Sites		**14 Points**
Y				Prereq 1	**Construction Activity Pollution Prevention**		Required
				Credit 1	**Site Selection**		1
				Credit 2	**Development Density & Community Connectivity**		1
				Credit 3	**Brownfield Redevelopment**		1
				Credit 4.1	**Alternative Transportation,** Public Transportation Access		1
				Credit 4.2	**Alternative Transportation,** Bicycle Storage & Changing Rooms		1
				Credit 4.3	**Alternative Transportation,** Low-Emitting & Fuel-Efficient Vehicles		1
				Credit 4.4	**Alternative Transportation,** Parking Capacity		1
				Credit 5.1	**Site Development,** Protect or Restore Habitat		1
				Credit 5.2	**Site Development,** Maximize Open Space		1
				Credit 6.1	**Stormwater Design,** Quantity Control		1
				Credit 6.2	**Stormwater Design,** Quality Control		1
				Credit 7.1	**Heat Island Effect,** Non-Roof		1
				Credit 7.2	**Heat Island Effect,** Roof		1
				Credit 8	**Light Pollution Reduction**		1

	Yes	?	No				Points
					Water Efficiency		**5 Points**
				Credit 1.1	**Water Efficient Landscaping,** Reduce by 50%		1
				Credit 1.2	**Water Efficient Landscaping,** No Potable Use or No Irrigation		1
				Credit 2	**Innovative Wastewater Technologies**		1
				Credit 3.1	**Water Use Reduction,** 20% Reduction		1
				Credit 3.2	**Water Use Reduction,** 30% Reduction		1

	Yes	?	No				Points
					Energy & Atmosphere		**17 Points**
Y				Prereq 1	**Fundamental Commissioning of the Building Energy Systems**		Required
Y				Prereq 2	**Minimum Energy Performance**		Required
Y				Prereq 3	**Fundamental Refrigerant Management**		Required

*Note for EAc1: All LEED for New Construction projects registered after June 26th, 2007 are required to achieve at least two (2) points under EAc1.

				Credit 1	**Optimize Energy Performance**		1 to 10
					10.5% New Buildings or 3.5% Existing Building Renovations		1
					14% New Buildings or 7% Existing Building Renovations		2
					17.5% New Buildings or 10.5% Existing Building Renovations		3
					21% New Buildings or 14% Existing Building Renovations		4
					24.5% New Buildings or 17.5% Existing Building Renovations		5
					28% New Buildings or 21% Existing Building Renovations		6
					31.5% New Buildings or 24.5% Existing Building Renovations		7
					35% New Buildings or 28% Existing Building Renovations		8
					38.5% New Buildings or 31.5% Existing Building Renovations		9
					42% New Buildings or 35% Existing Building Renovations		10
				Credit 2	**On-Site Renewable Energy**		1 to 3
					2.5% Renewable Energy		1
					7.5% Renewable Energy		2
					12.5% Renewable Energy		3
				Credit 3	**Enhanced Commissioning**		1
				Credit 4	**Enhanced Refrigerant Management**		1
				Credit 5	**Measurement & Verification**		1
				Credit 6	**Green Power**		1

continued...

	Yes	?	No				Points
					Materials & Resources		**13 Points**
Y				Prereq 1	**Storage & Collection of Recyclables**		Required
				Credit 1.1	**Building Reuse,** Maintain 75% of Existing Walls, Floors & Roof		1
				Credit 1.2	**Building Reuse,** Maintain 100% of Existing Walls, Floors & Roof		1
				Credit 1.3	**Building Reuse,** Maintain 50% of Interior Non-Structural Elements		1
				Credit 2.1	**Construction Waste Management,** Divert 50% from Disposal		1
				Credit 2.2	**Construction Waste Management,** Divert 75% from Disposal		1
				Credit 3.1	**Materials Reuse,** 5%		1
				Credit 3.2	**Materials Reuse,** 10%		1
				Credit 4.1	**Recycled Content,** 10% (post-consumer + ½ pre-consumer)		1
				Credit 4.2	**Recycled Content,** 20% (post-consumer + ½ pre-consumer)		1
				Credit 5.1	**Regional Materials,** 10% Extracted, Processed & Manufactured Regionally		1
				Credit 5.2	**Regional Materials,** 20% Extracted, Processed & Manufactured Regionally		1
				Credit 6	**Rapidly Renewable Materials**		1
				Credit 7	**Certified Wood**		1

	Yes	?	No				Points
					Indoor Environmental Quality		**15 Points**
Y				Prereq 1	**Minimum IAQ Performance**		Required
Y				Prereq 2	**Environmental Tobacco Smoke (ETS) Control**		Required
				Credit 1	**Outdoor Air Delivery Monitoring**		1
				Credit 2	**Increased Ventilation**		1
				Credit 3.1	**Construction IAQ Management Plan,** During Construction		1
				Credit 3.2	**Construction IAQ Management Plan,** Before Occupancy		1
				Credit 4.1	**Low-Emitting Materials,** Adhesives & Sealants		1
				Credit 4.2	**Low-Emitting Materials,** Paints & Coatings		1
				Credit 4.3	**Low-Emitting Materials,** Carpet Systems		1
				Credit 4.4	**Low-Emitting Materials,** Composite Wood & Agrifiber Products		1
				Credit 5	**Indoor Chemical & Pollutant Source Control**		1
				Credit 6.1	**Controllability of Systems,** Lighting		1
				Credit 6.2	**Controllability of Systems,** Thermal Comfort		1
				Credit 7.1	**Thermal Comfort,** Design		1
				Credit 7.2	**Thermal Comfort,** Verification		1
				Credit 8.1	**Daylight & Views,** Daylight 75% of Spaces		1
				Credit 8.2	**Daylight & Views,** Views for 90% of Spaces		1

	Yes	?	No				Points
					Innovation & Design Process		**5 Points**
				Credit 1.1	**Innovation in Design:** Provide Specific Title		1
				Credit 1.2	**Innovation in Design:** Provide Specific Title		1
				Credit 1.3	**Innovation in Design:** Provide Specific Title		1
				Credit 1.4	**Innovation in Design:** Provide Specific Title		1
				Credit 2	**LEED® Accredited Professional**		1

	Yes	?	No				Points
					Project Totals (pre-certification estimates)		**69 Points**

Certified: 26-32 points, **Silver:** 33-38 points, **Gold:** 39-51 points, **Platinum:** 52-69 points

ning a building. Instead of designing a building with the equipment one expects it to need, such as an air-conditioning unit, the designer takes into account the conditions, such as thermal comfort, the occupants would require once the building is opened. In this formulation, a mechanical engineer would set a target temperature for acceptable comfort. This typically follows the American Society of Heating, Refrigerating, and Air-conditioning Engineers' Standard 55, Thermal Environmental Conditions for Human Occupancy, which establishes a range of between 68 and 78 degrees Fahrenheit for acceptable human comfort. After a target temperature is determined, the mechanical engineer would then review the options that could be used to achieve the temperature. This could be a conventional system, such as a chiller and air-handling unit on the roof to supply ceiling-mounted ductwork and diffusers in the occupied space. Or it could instead include more exotic systems, like chilled beams, under-floor air distribution, automatically actuated operable windows, and exhaust ventilation that relies naturally on the stack effect. Ideally, integrated design would allow an investigation of all of these options, theoretically carrying them through to construction, in order to understand the most effective option for the building that would cause the least damage to the building. Conceivably, something like natural ventilation could not be considered outside of an integrated design approach, since the scheme relies on so many building systems and components that fall outside the traditional realm of the mechanical engineer.

While the consistency of the design process fluctuates with any project, almost every one of the consultants who effectively introduced integrated design to their method of working stresses the importance of setting goals early in the process. They say that getting the architect, engineers, contractors, consultants, and, perhaps most important, the client to agree to a set of design objectives needed to realize a project's potential can prove to be the difference between success and failure in green building.

THE U.S. GREEN BUILDING COUNCIL (USGBC) AND LEED

Recognizing the need for cross-collaboration to support the integrated design and realization of high-performance buildings, the USGBC formed in 1993 to establish what has become the Leadership in Energy and Environmental Design program, or the LEED Green Building Rating System. The USGBC was organized originally as a consensus-driven committee of like-minded people. They recognized there was a need both to educate the profession about how to create sustainable buildings and to teach owners and users of buildings about the value of green design.

This committee of professionals was drawn from a broad swath of the design and construction industry—architects, engineers, contractors, consultants, and

peripheral figures. They began to compile information about how buildings influence the environment, strategize ways in which new buildings could be evaluated prior to construction, and, in the words of the LEED reference guide, "define and measure 'green' buildings." The ultimate goal of LEED's creators was to effect a market transformation. They understood that for green building to become viable, their clients would have to understand what advantages building green would have for them in terms of life-cycle costs, productivity increases, and their ability to market their buildings. A green building would need to cost the same as or less than a conventional one, and they would need to be measurable against a commonly understood set of benchmarks tied to existing codes and standards.

What developed and was formally released in 2000 was a points-based system of guidelines for new construction (LEED-NC) that stretched across six categories: sustainable sites (14 points), water efficiency (5 points), energy and atmosphere (17 points), materials and resources (13 points), indoor environmental quality (15 points), and innovation and design process (5 points). Each category has sub-categories that are assigned a number of points, adding up to an overall possible score of 69 points for a LEED Platinum project. Other categories include Certification (26–32 points), Silver (33–38 points), and Gold (39–51 points). Many of the case studies featured in *GreenSource* have been certified at one of these levels, while others may have been designed to meet the requirements of a level without having actually been certified. We have made every attempt to verify the certification level if the project was awarded the distinction prior to this book's publication date. A LEED points chart is included with the case studies that have been certified. A typical mistake is to consider LEED a building standard, as if it prescribes certain technologies and methods for designing a green building. LEED is merely a ratings system that depends on existing standards. To go back to the ventilation example from our earlier discussion of integrated design, LEED may reference ASHRAE Standard 55, but it does not establish occupant comfort levels of its own. Thus, if ASHRAE were to change its standard, LEED would probably follow. The only way to change LEED's requirements for occupant comfort would then be through either the industry-consensus standards-developing process of ASHRAE or the industry-consensus process of LEED. This prevents a single group or stakeholder from altering LEED's requirements to its own benefit.

Since the successful deployment of LEED-NC, the USGBC has released similar ratings programs for commercial interior/tenant improvements, existing buildings/operation and maintenance, core and shell, and homes. These are structured like LEED-NC but include points and standards more specific to these individual markets. As of press time, LEED programs for neighborhood development, retail, and health care were in pilot phases. As of December 2007, there were 9,000 projects registered with the USGBC as attempting LEED certification in one of its programs, with an additional approximately 1,200 projects completed and certified throughout all 50 states and in 41 countries. By any measure, a LEED rating on a building has become a signifier of achievement in the design and construction industry.

But the overall rating is often less interesting than the specific ways the design and construction team achieved that rating, which brings us to a brief discussion of the details of each category. In LEED for New Construction version 2.2, which is the latest available iteration of LEED for new buildings, the sustainable sites (SS) category has eight credits and one prerequisite. The prerequisite, for construction-activity pollution prevention, is required in order to even consider certifying one's building under the LEED program. Each category has one to three prerequisites, but these rarely hinder a project. In the case of sustainable sites, the prerequisite ensures that construction waste, erosion, and runoff do not contaminate adjacent sites or the overall environment. A project does not get points for a prerequisite. The points in the sustainable-sites category include alternative transportation (such as access to public transportation), brownfield redevelopment (cleaning up a contaminated site, as opposed to using a "greenfield," which has never been built on), light pollution reduction, and mitigation of the heat-island effect.

There is no one way to earn these points, as the LEED committee did not want to prevent designers from providing innovative, unconventional solutions to problems that often depend on the specific conditions of each site and project program. For example, with KieranTimberlake Associates' Sidwell Friends Middle School in Washington, D.C., the architects achieved a point for SS Credit 6 by installing a constructed wetland, the first of its kind inside the District of Columbia and somewhat of a test case for using the technology in a dense urban environment. In part by embracing what

is arguably a radical idea for a children's school, in addition to comprehensive sustainable strategies throughout the project, KieranTimberlake helped the Sidwell project achieve a LEED Platinum rating. But the point remains, LEED does not demand the installation of a constructed wetland to achieve SS Credit 6.

The other credits work similarly. The water-efficiency credits look at ways to handle wastewater, the reduction of irrigation for landscaping, and water-use reduction. Energy and atmosphere requires some fundamental building commissioning, minimum energy performance, and reduction of chlorofluorocarbons in mechanical systems, but it also encourages the use of renewable-energy strategies, more comprehensive commissioning, the installation of measurement and verification devices on building systems, and the purchase of so-called green power.

The materials and resources credits require the storage and collection of recyclables, as well as encourage building reuse, recycled content of materials, and separate credits for the use of certified wood, local or regional materials, and rapidly renewable materials. These credits merit a bit more consideration, as one of the chief struggles of the green building designer is gauging what materials necessarily qualify as "green." The USGBC does not certify or label materials as being green, but it does recognize a few certifications like that of the Forest Stewardship Council (FSC), which certifies that wood has been harvested from responsibly managed forests. There are many others, like the Master Painter's Institute's Green Performance standard, which is recognized by LEED in order to comply with indoor-air-quality requirements for low-volatile-organic compounds (VOCs) in finishes, as well as furniture and products. But green product certifications and labels such as these are subject to constant change and many are produced by manufacturers and trade organizations that are obviously promoting an agenda. It is best to conduct your own background research on a product's green performance claims or to consult with a trusted industry resource.

Leed Scores

All of the case studies in *Emerald Architecture* that are LEED certified include a LEED scores chart. This enables readers to understand which aspects of sustainability were emphasized during the building's design. For example, the scorecard at right indicates that only six of a possible 17 points for "energy" were earned, which may indicate the building's energy performance could be improved. But, it achieved five of five points for "innovation," which shows that the designers may have devised some unusual means to make the structure more environmentally sound than it would have been otherwise.

LEED SCORES (BUILDING A/PHASE 1)
LEED-NC Version 2 Gold

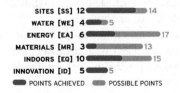

	Achieved	Possible
SITES [SS]	12	14
WATER [WE]	4	5
ENERGY [EA]	6	17
MATERIALS [MR]	3	13
INDOORS [EQ]	10	15
INNOVATION [ID]	5	5

■ POINTS ACHIEVED   ■ POSSIBLE POINTS

Key Parameters

> In order to make the case studies in this book more meaningful, each includes important facts and figures, such as the building site's geographic location. Cost-per-square-foot figures answer the question that is probably asked more frequently than any other during discussions of architecture. However, how much energy is purchased annually and how much CO_2 is at the source of it are important metrics when considering how efficient a building may be. These figures are sometimes based on predictions, and sometimes based on actual metrics: the word "predicted" appears when they are predicted and "actual" when actual.

KEY PARAMETERS

LOCATION: Tempe, Arizona (Salt River watershed)

GROSS SQUARE FOOTAGE: 347,000 ft² (32,240 m²)

COST: $104 million

COMPLETED: December 2004 (Building A); November 2005 (Building B)

ANNUAL PURCHASED ENERGY USE (BASED ON SIMULATION): 145 kBtu/ft² (1,650 MJ/m²),

ANNUAL CARBON FOOTPRINT: (predicted): 19.1 lbs. CO_2/ft² (93.3 kg CO_2/m²)

PROGRAM: Laboratory space, lab support spaces, open office space, private offices, conference rooms, atrium (collaborative space), public entry lobby, auditorium, and café.

The indoor environmental quality credits require that buildings be non-smoking environments and meet industry-accepted ASHRAE standards for indoor-air quality. The optional credits include carbon dioxide monitoring, the use of low-VOC materials (as stated previously in regard to paint), increased occupant controllability of mechanical systems, greater occupant thermal comfort, and access to daylight and views. Finally, the innovation and design-process credits, which are all entirely optional, allow the design team to submit innovative design schemes to the USGBC for individual review. This is where a team that adopts a radical technology could achieve additional LEED points.

All of the information needed for complying with a given level of LEED certification must be collected throughout the design process, formatted to the specifications of the USGBC, and submitted to the USGBC's LEED reviewers for evaluation after the building has been opened. Once the material has been successfully assessed by the USGBC and deemed acceptable, a certification level is assigned, thus rewarding the building with a LEED rating. Usually engraved on a plaque, the LEED rating is typically displayed near the entrance of a building. This process sounds easy on paper, but for a large project going after many points, it can necessitate the full-time attention of a knowledgeable staff member for months at a time. A cottage industry of LEED consultants who specialize in preparing this paperwork has sprung up to address this problem; many times, a project team will decide against pursuing an official LEED rating strictly due to the cost of completing and filing this documentation. The USGBC has streamlined the process considerably in the last few years, including incentives for projects pursuing LEED Platinum, but it continues to be labor-intensive in 2008.

While the general consensus of the design and construction industry is that the USGBC and LEED have been positive change agents for the market's embrace of green buildings, there has been a backlash among some designers who see the widespread adoption of LEED as the reduction of design quality to a checklist of points. The editors at *GreenSource* have been willful about selecting projects, both LEED-certified and not, that demonstrate true sustainability concerns in at least one or more aspects of their design and construction. This approach eschews the projects that merely aim for LEED points without taking into consideration that the best architecture exceeds its users' expectations, stimulates the imagination, and causes the human spirit to soar. No one has ever been able to devise any checklist that, if followed during design, can guarantee inspirational architecture will result. That is a good reason why the pages of this book should be studied as much for what they do not include as for what they do.

Additionally, some critics of LEED argue the process of developing the ratings system is too slow and cumbersome to address the dramatic and changing problems of global climate change. Others argue the LEED system itself is flawed, giving as many points for adding bike racks and showers to a project as it does for shaving 10.5 percent off its overall energy use. While there is something to be said for those critiques, one must remember LEED was never meant to hold back the ambitions of the most eco-conscious designers but to transform the conventional building marketplace across a broad spectrum of design issues. This is still a challenge, as the number of LEED-rated buildings in the U.S. is relatively paltry when compared to what gets built every year. McGraw-Hill Construction Analytics estimated that in 2007, green building projects represented nearly $12 billion, which is small compared to a construction market of around $1.4 trillion. Still, the continuing adoption of the LEED program by states, municipalities, and federal agencies seems to ensure the program will become a much more widespread requirement for new buildings in the next several years. On some level, the growing membership of the USGBC assures this, since nearly 91,000 individuals have joined, along with almost 13,000 organizations. The editors of this book trust the case studies presented here will inspire countless others to pursue action.

THE *GREENSOURCE* CASE STUDIES

The editors of *Emerald Architecture* also hope that the case studies chosen from *GreenSource* will form a collection of examples of the very best of green design in the U.S., with a few foreign projects also included, that can be used to educate readers—architects, engineers, contractors, interior designers, clients, product manufacturers, financial industry leaders—in the contemporary methods of sustainable design. Each case study takes a different editorial tack, focusing on what the editors feel is the strongest contribution of the project toward sustainable-design knowledge. Sometimes that can mean the integrated-design process that leads to the project's success or the social component of a building's community base, while other times it may be the difficulty of using untested technology or the realities of market-rate construction.

The editors want to make these case studies more useful than a simple one-to-one comparison of LEED ratings, unlike other publications and approaches. Thus, a set of key parameters has been included with each one. These include the annual purchased energy use and the annual carbon footprint that energy represents. As more of these case studies are written, energy-use and carbon-footprint numbers could be gathered and compared in order to begin to form benchmarks that could help us further understand how the design of our buildings affects our environment. But how does *GreenSource* establish these numbers?

As many in the sustainable design industry already know, there are no industry standards for determining the carbon footprint of a building. In fact, the issue of carbon footprints for buildings did not really take hold in the design and construction industry until 2007; indeed, LEED does not take into account a carbon footprint. However, that is likely to change in the future. In the meantime, *GreenSource's* editors rely on a crude approximation based on energy bills provided by the owner of the building profiled or on the predicted energy use from an energy model provided by a building's environmental consultant. From one of these sources, an estimate of carbon dioxide emissions is determined using the U.S. Environmental Protection Agency's online Power Profiler (www.epa.gov/cleanrgy/powerprofiler.htm). However, this only considers emissions tied to electricity. For other energy sources, the editors turn to factors derived from the Greenhouse Gas Protocol Initiative (www.ghgprotocol.org), which also takes into account regional energy use and production. Together, an estimate is derived and that is what the editors share with readers. It's a non-scientific approach to be sure, but

the editors see it as the beginning of a process of documentation that cannot afford to wait for a larger industry consensus.

Also included with each case study are three charts for sky conditions, temperatures and dew points, and heating-and-cooling degree days. These charts are provided as a way to flesh out the environmental context of each project, so readers can get a sense of the natural conditions that influenced each design decision. The sky conditions give us an idea of the percentage of days throughout the year that are cloudy, mixed, or clear. For example, it would follow that a building in a climate where the percentage of clear days exceeds 50 percent would obviously necessitate the design of a shading system or another device that would mitigate such frequent sunshine. The temperature and dew-point chart reflects the range of annual temperatures broken down by month, including the average temperature year-round.

A high-relative-humidity climate, where the average temperature regularly and substantially exceeds the dew point, could, for example, lead to design options like using condensation from a building's mechanical system to flush toilets. The heating-and-cooling degree-days chart illustrates the demands on the mechanical system to provide heating or cooling, sometimes both, in any given month. This could explain why some buildings, no matter how many passive ventilation schemes are designed into them, still rely on conventional mechanical systems to offset harsh conditions. It is also worth noting that the key parameters sections also indicate where the watershed in each project is located.

Finally, like in most McGraw-Hill Construction publications, the editors of *GreenSource* provided the names of manufacturers of key building materials and systems, and credit members of the project teams. Manufacturers' names are meant to aid other designers in finding materials and systems that could be used in realizing their own green building projects. It should be noted, however, that, following the previous discussion of materials, it is up to every designer to do whatever research may be required to determine whether the selections they make are green ones.

Emerald Architecture's case studies are arranged in six chapters according to project type: civic and cultural, education, government, offices, residential, and science and technology. A short essay introduces each chapter and explains the challenges faced by the designers of that specific building type, particularly when they applied principles of sustainability to it. The editors expect these case studies to enlighten and engage, as well as provoke, but, ultimately, to lead to more green buildings and a better built environment for everyone. **«**

Russel Fortmeyer and Charles Linn, FAIA

Chapter 2

CIVIC/CULTURAL

CIVIC AND CULTURAL CENTERS HAVE GREAT POTENTIAL TO WORK AS CITADELS of environmentally conscientious behavior. By garnering public interest and educating employees and visitors to the benefits of green design features, these buildings can promote the cause of sustainable living to the average person. Furthermore, eco-conscious design makes good business sense for these primarily non-profit organizations: By incorporating energy-efficient and passive systems into their buildings, they can reduce operational costs, which can bog down daily operations over time. The following four case studies are examples of civic and cultural buildings that stand out for their approach to green design.

Almost across the board, visitation has increased for the four public buildings discussed in this section. In addition to the usual sustainable features like radiant heating and cooling or vegetated roofs, many of these buildings incorporate a public display of the sustainable features at work. Seventy percent of the electricity needs for the Water + Life Museums in Hemet, California, are met with 50,000 square feet of solar photovoltaic panels artfully installed atop both buildings. Although the photovoltaics were initially a controversial decision, Darcy Burke, the executive director of the Water Center, says an exhibit showing how the building produces and consumes energy has been a great hit with visitors.

The Central Library in Des Moines, Iowa, employs an expanded-copper fine mesh laminated into triple-glazed panels to form the thermal envelope for the building. This mesh shades the building during the day and, from the outside, fades from copper to transparent when the sun sets. This compelling visual effect broadcasts to Des Moines the beauty of this sustainable feature and, more generally, the library's finely calibrated relation to its environment. Workers are ecstatic and visitors charmed with New York's light-filled, LEED Silver Bronx Library Center, which has become a de facto community center for a thriving neighborhood. The renovation and addition to the Toronto Botanical Gardens, also LEED Silver, incorporates demolished material into its fourteen different gardens, which feature a variety of exotic plants in organic soil. Something to take away from these four projects is their ability to foreground their sustainability while remaining aesthetically pleasing to a variety of community members, making public awareness a goal to be joined to the efficient use of energy and resources. ◀◀

Green Thumb

A NEW, LIGHT-FILLED ENTRANCE AND ADDITION ESTABLISHES A
BOTANICAL GARDEN'S REPUTATION

GEORGE AND KATHY DEMBROSKI
CENTRE FOR HORTICULTURE
TORONTO BOTANICAL GARDEN

NADAV MALIN

T O TORONTO RESIDENTS WHO WALK INTO THE TORONTO BOTANICAL Garden's new facilities, it seems obvious now but when David Sisam, principal of Montgomery Sisam Architects, suggested eliminating a series of entrance ramps and converting what was a split-level design into a simple, grade-level opening, it was a revelation. Previously, visitors to the facility walked up a berm from the parking area and entered the facility between the ground floor and second floor. Inside, large ramps led up or down. By removing both the berm and the ramps, Sisam found new space and simplified the layout. That intervention structured the design of the renovation and addition to two buildings in an expansion that transformed the sleepy Civic Garden Centre into a major urban botanical garden. The existing site conditions included administrative offices in a 1964 wood-and-stone building designed by Raymond Moriyama, connected to a larger conference and meeting facility built in

LEED SCORES
LEED Canada NC 1.0 Silver

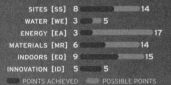

	Points	Possible
SITES [SS]	8	14
WATER [WE]	3	5
ENERGY [EA]	3	17
MATERIALS [MR]	6	14
INDOORS [EQ]	9	15
INNOVATION [ID]	5	5

POINTS ACHIEVED   POSSIBLE POINTS

KEY PARAMETERS

LOCATION: Toronto, Ontario, Canada (Don River watershed)

GROSS SQUARE FOOTAGE: 3,800 ft² (353 m²) addition and 4,300 ft² (400 m²) renovation

COST: $3.25 million (buildings) and $4.3 million (landscape and gardens)

COMPLETED: December 2005

ANNUAL PURCHASED ENERGY USE (BASED ON SIMULATION): 80 kBtu/ft²(912 MJ/m²), 30% reduction from base case

ANNUAL CARBON FOOTPRINT (PREDICTED): 12 lbs. CO_2/ft² (57 kg CO_2/m²)

PROGRAM: Library, administrative offices, store, children's center, and meeting areas

TEAM:

OWNER: City of Toronto

ARCHITECT: Montgomery Sisam Architects

LANDSCAPE: PMA Landscape Architects and Thomas Sparling

ENGINEERS: Blackwell Bowick Partnership (structural); Rybka Smith and Ginsler (mechanical & electrical);

ENVIRONMENTAL, ENERGY, AND COMMISSIONING: Enermodal Engineering

GENERAL CONTRACTOR: The Dalton Company

SIGNAGE: Adams + Associates

TOM ARBAN PHOTOGRAPHY

1976 and designed by Jerome Markson. These buildings, as venerable as they were, could not adequately satisfy the Centre's ambitions for expanding its programming. The design team's solution was to reconfigure the entrance lobby of the Markson building, removing problematic south-facing sloped glazing, and to add a new 3,800-square-foot retail wing. The addition, which sports a sloped, vegetated roof, clear glazing at ground level, and translucent panels above, helps define several new outdoor spaces. "The design is all about providing a sense of the connection between the indoors and outdoors. We joked that usually architects take the lead on projects like this, but here the landscape architects took charge because we all acknowledged that the gardens are the most important part," notes Sisam.

Owned by the City of Toronto, the facilities are located at one end of Edwards Park, a large estate-turned-public-park that is part of a greenway linking open spaces throughout the city. A $350,000 challenge grant from the Kresge Foundation launched the organization's fundraising campaign. Kresge offered an additional $150,000 if the project was LEED-certified, which executive director Margo Welsh says was an easy decision for the ecologically oriented group. With a total budget of nearly $7.6 million dollars, $3.3 million went to the renovation and addition. The remainder covered the landscape work, design fees, and other soft costs.

Two firms, PMA Landscape Architects and Thomas Sparling, collaborated to create a landscape surrounding the new building that includes 14 different gardens on less than four acres of property, according to PMA's Jim Melvin. With the exception of the entry garden, which was commissioned by the Garden Club of Toronto, the two firms designed all the gardens and walkways on the site. While some of the gardens feature exotic plants, they were installed using only organic soils and amendments. The designers focused on utilizing material from the site and working the reuse of demolished material into their design. "We inventoried all the material before demolition started, so we knew exactly what we had to work with," Melvin says, adding that "there was a bit of a battle between us and the architect as to who would get the better stone."

The site sports two significant water features, both of which recirculate water internally for a week before draining to underground storage tanks for irrigation use. By replacing the water weekly, the Botanical Garden avoids having to add chemicals, and the water doesn't concentrate to the point where mineral deposits mar the surfaces, according to Melvin. The municipal water used to replenish the water features is the only use of potable water in the landscape, as all the irrigation is supplied from reclaimed water. In addition to

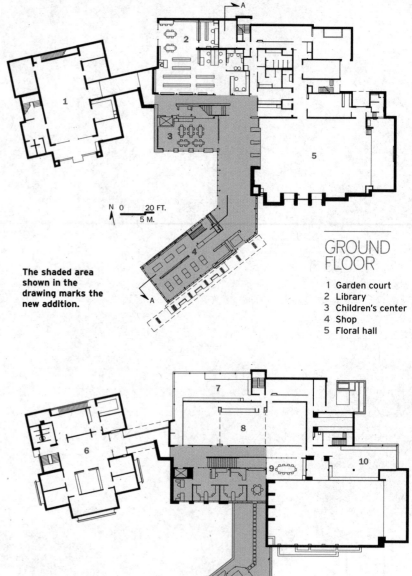

The shaded area shown in the drawing marks the new addition.

GROUND FLOOR

1 Garden court
2 Library
3 Children's center
4 Shop
5 Floral hall

SECOND FLOOR

6 Administration
7 Garden club
8 Studios
9 Meeting room
10 Milne house
11 Green roof

« A rooftop patio in the renovated building overlooks the entry plaza, with the new addition at right.

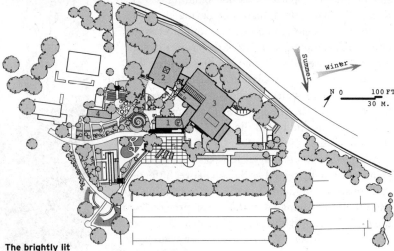

The brightly lit interior spaces (left top and center) provide a visitor experience that lives up to the high standards of the gardens outside, which are punctuated by views of the sloping vegetated roof (right top) and finely wrought fence details (left).

SITE PLAN

1 New addition
2 Existing building from 1964
3 Existing building from 1976
4 Greenhouse
6 Arrival courtyard

the water features, roof runoff and condensation from the air conditioners feed irrigation tanks. Channels lead stormwater from the hardscape areas to underground infiltration pipes, where it irrigates deep-rooted plants before percolating into the soil. Runoff from the parking area goes to city drains, however. "When the parking lot gets redone, I hope they will use a permeable paver," suggests Melvin.

Both the addition and the renovated space are infused with daylight that, together with direct views, helps bridge the exterior landscape and the interior spaces. "During the day, very rarely are lights on inside," says Sisam. Environmental design and energy consultants Enermodal Engineering supported that work with energy and daylight modeling in software program Ecotect. "This gave us a very quick analysis and understanding of what was going to happen in the spaces," says Enermodal's Braden Kurczak. Supported by the simulations, Kurczak was able to convince the designers to use an opaque wall surface up to 18 inches from the floor. "The architects were all about having floor-to-ceiling glass, but there's not much benefit in terms of light from any glazing below your knees," Kurczak notes. The effect of this change on the building's thermal performance was not huge, but it was one of many little things that added up to make a big difference, according to Kurczak.

In addition to the sill, the clear glass on the south side of the retail space in the new building is shaded by a trellis, while the translucent glazing above is fritted to reduce solar gain and glare from direct sun. Other energy-saving features of the project include efficient and well-distributed lighting fixtures, demand-controlled ventilation, and, for the new retail area, an enthalpy wheel providing energy-recovery in the ventilation air. The firm RS&G Commissioning reviewed plans and tested the installation to ensure that it operates as intended.

As construction manager, Rick Gosine of The Dalton Company served as liaison between the design and construction teams and the commissioning agent. Dalton also participated in the design process from the beginning, providing cost estimates on the various design alternatives. "When the design was exceeding our price expectations, we would come up with a plan to bring it back on budget," Gosine says. Kurczak credits Dalton with coming up with an inexpensive solution for erosion control by putting silt fencing right up against the base of the eight-foot plywood hoarding wall that surrounded the construction site. Kuczak notes that the usual approach, using stand-alone silt fencing, would have required digging up paving in the parking lot.

For the purposes of the LEED submission, the scope of the project was limited to the 3,800-square-foot addition and a 4,300-square-foot area of substantial

« A trellis shades the clear vision glass of the new addition's retail space, while softening the transition between building and gardens.

« Above the trellis, fritted glass filters and diffuses sunlight, creating an indoor space that is bright while minimizing glare and heat gain.

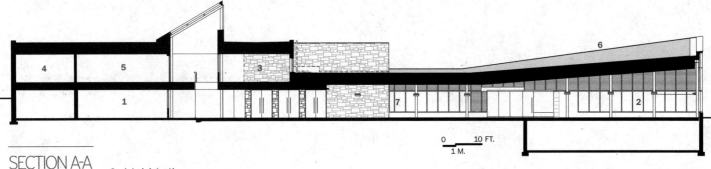

SECTION A-A

1 Library
2 Shop
3 Administration
4 Garden club
5 Studios
6 Green roof
7 Main lobby

0   10 FT.
1 M.

renovation, according to Kurzcak. The predicted 26 percent energy cost savings were enough to earn three energy optimization points in a renovation project, but only one in new construction. Given its nearly even split of new construction and renovation, the project earned two points.

Only minor steps were taken to improve the energy efficiency of the remaining 28,200 square feet of the existing building, so the facility's overall energy use remains high. "Mechanical and electrical systems really do need upgrading in the existing buildings," acknowledges Sisam. "When we were doing renovations, we found that we had to do more upgrading than we expected, because every time we opened something up it would fall apart," he says. Welsh points out that it's harder to raise money for improvements to existing facilities than for a new building. "We would like to have had more money to make greater improvements in the old buildings," she says.

The ongoing need to address the older buildings aside, Welsh is thrilled with the project. "The building is beautiful, and the fact that it is certified LEED Silver excites people," she says. Welsh is especially pleased that many city departments are choosing to hold meetings in the renovated facility, giving them a first-hand experience of a successful LEED project. The project demonstrates that "you can have something that is great to be in, looks great, and is environmentally responsible," she says. ❮❮

SKY CONDITIONS

The frequently cloudy skies make a brightly lit indoor space all the more appealing.

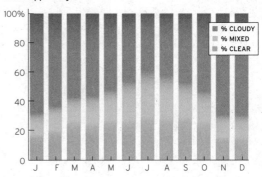

TEMPERATURES

In cold, the transparent walls of the addition allow visitors a way to experience the gardens from the indoor heated space.

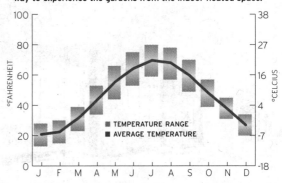

HEATING/COOLING DEGREE DAYS

Heating is the dominant demand on the mechanical system in this cold-climate facility.

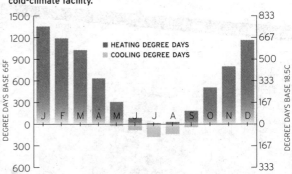

SOURCES

GLASS: Prel-Coat Ceramic frit coated glass - Prelco Inc.

GREEN ROOF ASSEMBLY: Sopranature by Soprema

PAINTS AND STAINS: Devthane 389; Glidden 36600, 94410, 94500, 94900; Selectone M99 Primer; Allcolour Lead Free Primer

FLOOR AND WALL TILE: Natural Vermont Slate - Sheldon Slate

CARPET: Interface Flooring Systems - Earth Carpet Tile

DOWNLIGHTS: Portfolio M6023T; Portfolio M7210T

EXTERIOR LIGHTING: Bega 7460; Erco Axis LED Walklight; Erco Beamer II Projector

A Desert Oasis

TWO NONPROFITS JOIN TO CREATE PUBLIC EXHIBITIONS AND
OFFICES DEMONSTRATING THEIR COMMITMENT TO SUSTAINABILITY

The roofs of the Water and Life museums
are nearly all covered with solar panels.
Although hidden from view for visitors
standing near the buildings, the panels are
visible from a distance.

WATER + LIFE MUSEUMS
HEMET, CALIFORNIA

JESSICA BOEHLAND

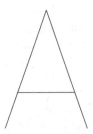

NGLER MIKE LONG CAUGHT A 16.4-lb. largemouth bass while fishing on Diamond Valley Lake, California, in March 2007, setting a new lake record. Amazing, considering that just two decades ago, this 4,500-acre lake was farmland in the saddle between two mountain ranges. Three dams, 260 billion gallons of water, and $2 billion later, however, the reservoir could meet Southern California's water needs for six months in the event of an emergency.

Located about 300 feet below the eastern dam, two new museums are part of a multimillion-dollar thank-you gift from the Metropolitan Water District (MWD) of Southern California to the community of Hemet for allowing the reservoir—the largest earthworks project in U.S. history—to be built in its backyard. "Water politics is big and complex," says Michael Lehrer, FAIA, principal of Lehrer + Gangi Design + Build, with just a hint of understatement. The Western Center for Archaeology and Paleontology displays fossils and Native American artifacts unearthed during excavation for the dams, while the Center for Water Education teaches visitors the importance of water in Southern California and its impact on the rest of the world.

Lehrer and Mark Gangi, AIA, who together led the design process, intended the project to honor the architectural tradition of large infrastructure projects, perhaps recalling the power and precision of the turbines in the dam. The architects wanted something "Stonehenge-ian," according to Lehrer, with abstract, geometric volumes jutting out from the landscape. The resulting buildings intersperse large window walls with steel-clad towers, creating a modern, industrial aesthetic.

The 62,000-square-foot project houses facilities for both centers, including exhibit and interactive space, laboratories, classrooms, offices, a gift shop, and a café in its multiple buildings. While each of the two main buildings has its own air handler and radiant floor manifolds, they share a boiler, a chiller, and a building

KEY PARAMETERS

LOCATION: Hemet, Calif. (Domenigoni Valley no watershed)

GROSS SQUARE FOOTAGE: 62,215 ft² (5,780 m²)

COST: $36.8 million

COMPLETED: November 2006

ANNUAL PURCHASED ENERGY USE (BASED ON SIMULATION): 20.3 kBtu/ft² (230 MJ/m²)

ANNUAL CARBON FOOTPRINT (PREDICTED): 4 lbs. CO_2/ft² (21 kg CO_2/m²)

PROGRAM: Exhibits, administration, meeting rooms

TEAM

OWNER: The Center for Water Education, Western Center for Archaeology and Paleontology

ARCHITECT: Lehrer + Gangi Design + Build

LANDSCAPE: Mia Lehrer and Associates

ENGINEERS: Nabih Youssef & Associates (structural); KPFF Consulting Engineers (civil); IBE Consulting Engineers (MEP/commissioning); Vector Delta Design Group (electrical/solar)

LEED CONSULTANT: Zinner Consultants

EXHIBIT DESIGNERS: Design Craftsmen

COMMISSIONING AGENT: IBE Consulting Engineers

BENNY CHAN

management system that controls and monitors the mechanical systems.

The buildings' eastern orientation, along a road that leads to a marina, posed a challenge for passive solar design. In response, the design team extended the project's prominent towers 16 feet beyond the glazing, shading the glass from direct sun at all times except early morning. Energy simulations revealed that even morning sun would cause unacceptable heat gain, however, as the mullions continued to radiate. Seeking shade but worried about the potential for dust and sand to compromise a mechanized system, the team made the potentially controversial choice to install large disposable scrims, sort of like exterior curtains. "We kept the banners eight feet from the ground, so from inside there's still a clear view out to the valley," says Gangi. The PVC-coated polyester banners, which hang across 10,000 square feet of glass, will last only three to five years, but Gangi says they're easy to replace.

Neither museum's board of directors was originally interested in green design. After nine months of explaining and modeling its benefits, however—especially its potential to reduce operational costs, an endless headache for nonprofit organizations—the design team convinced both boards to embrace green design. The change of heart led to significant design modifications. "We went from air-handling units to radiant heating and cooling, which meant redesigning the entire slab," says Lehrer. Since the system adjusts the temperature at the floor, where occupants and exhibits are, instead of at the ceiling, it saves considerable energy. The buildings' high ceilings—32 feet in the front space—made this decision especially beneficial.

It took another nine months for the design team to convince the Water Education Center of the benefits of photovoltaics. Gangi says he originally thought it was a stretch to expect a water museum to invest in solar power, "but then we learned that MWD is Southern California Edison's [the energy supplier] largest customer." The decision to install a solar array came down to a fateful meeting between the design team and Phillip Pace, then the director of MWD and chair of the water center. "I insisted that if they didn't do this, they'd come back in two years and say, 'What the hell were you thinking by allowing us to proceed without this?'" recalls Lehrer.

The ensuing photovoltaic array, which covers 50,000 square feet atop almost all of both buildings, currently generates nearly 70 percent of the project's electricity needs, according to Peter Gevorkian, of Vector Delta Design Group, Glendale, California, who designed the system. The 540-kW installation, using 185-watt modules from Sharp, was built by electrical subcontractor Morrow-Meadows Industry, California. Although the system cost $4 million, rebates from the California Energy Commission and Southern California Edison's Savings by Design program cut the price in half, yielding an anticipated seven-year payback.

The original LEED goal was a certified rating. Once the photovoltaic system had been approved, however, the team set its sights on Silver and then Gold. John Zinner, the project's sustainable develop-

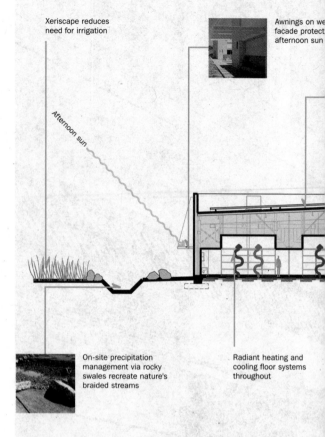

Xeriscape reduces need for irrigation

Awnings on we facade protect afternoon sun

Afternoon sun

On-site precipitation management via rocky swales recreate nature's braided streams

Radiant heating and cooling floor systems throughout

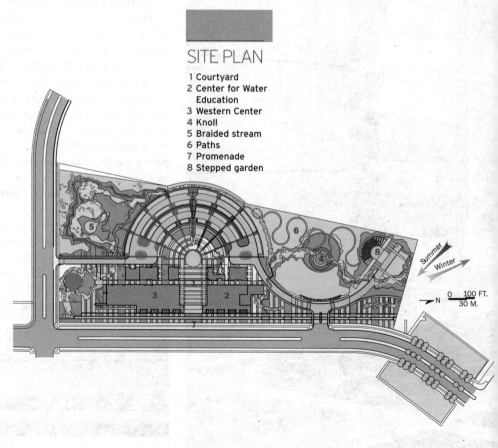

SITE PLAN

1 Courtyard
2 Center for Water Education
3 Western Center
4 Knoll
5 Braided stream
6 Paths
7 Promenade
8 Stepped garden

Summer
Winter

0   100 FT.
30 M.
N

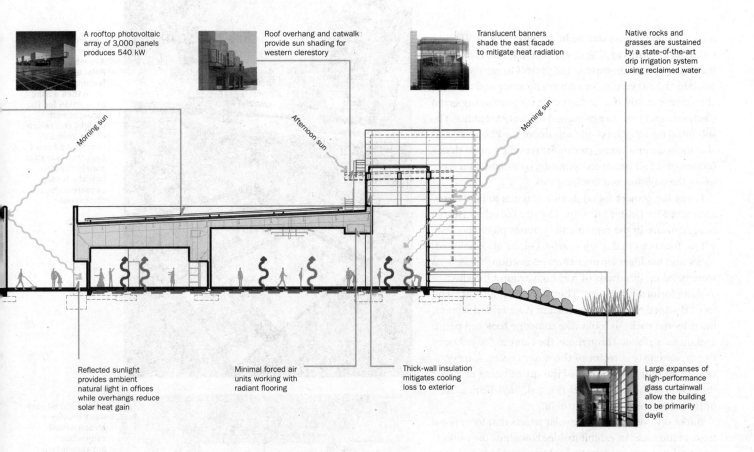

A rooftop photovoltaic array of 3,000 panels produces 540 kW

Roof overhang and catwalk provide sun shading for western clerestory

Translucent banners shade the east facade to mitigate heat radiation

Native rocks and grasses are sustained by a state-of-the-art drip irrigation system using reclaimed water

Morning sun

Afternoon sun

Morning sun

Reflected sunlight provides ambient natural light in offices while overhangs reduce solar heat gain

Minimal forced air units working with radiant flooring

Thick-wall insulation mitigates cooling loss to exterior

Large expanses of high-performance glass curtainwall allow the building to be primarily daylit

« Walkways leading to each museum's entrance are shaded by translucent solar panels overhead (opposite). The buildings feature large window walls interspersed with metallic monoliths in "a repeating pattern that helps your eye mark off the distance," according to architect Mark Gangi.

ment consultant, says that as he reviewed the LEED sub-mittal just months ago, "it dawned on me that it might be Platinum." The team expects the project to earn all of the available LEED credits for energy efficiency and renewable energy in addition to the credit for purchasing green electricity and two energy-related innovation credits. The submittal for 52 points—the minimum for Platinum—also includes innovation credits for recycling more than 95 percent of all construction waste, by weight, and for using the building as a teaching tool.

Using the project for education extends to landscape architect Mia Lehrer's design. The site tells the history of agriculture in the region and includes plant species whose fossils were dug up nearby. Lehrer also included rocks and boulders leftover from excavation. "They were piled up hundreds of feet high around the site, looking forlorn," she says. "They were magnificent." The most distinctive landscape element is a braided stream, lined by red rock. "It looks like someone took red paint and made a ribbon throughout the campus," says Darcy Burke, executive director of the water center. A recent rainstorm inundated much of the surrounding area, "but none of the campus was flooded," says Burke. "All of the water went into the stream."

Burke says visitors love the solar panels that form portions of the roof. An exhibit inside shows how the project uses and generates energy "in real time and in 'people speak,'" says Burke, "so you don't have to be an engineer to understand it." Dual-flush toilets, waterless urinals, and a drip-irrigation system that uses utility-supplied gray water in place of potable water teaches about water efficiency. The team selected interior furnishings for recycled content, low-chemical emissions, and regional availability. Much of the wood used meets Forest Stewardship Council standards for responsible harvesting.

The Western Center has been open since October 2006, and general admission has been running 20 percent to 30 percent above projections, according to executive director Bill Marshall. The Center for Water Education, on the other hand, has been struggling. Although it has hosted more than 2,500 visitors, the MWD took over the financially ailing nonprofit group operating the center and closed its doors.

While the project's upfront cost of $37 million hasn't helped the Water Center's current financial situation, nobody seems to blame the green design. To the contrary, Michael Lehrer believes "the fact that this is an environmental showcase will probably be its saving grace. It will end up garnering interest and support that it otherwise wouldn't have." Pace, who has taken heat for the financial shortfall, says he's received only positive feedback about the project's environmental responsibility. "Everyone understands we need to protect what we have," he says, "or we won't have it very long." And true to his thinking, the center reopened September 2007. ◀◀

◀◀ The Western Center for Archeology and Paleontology features fossils unearthed during excavation for the dams, in both upright recreations of the animals' forms, and buried beneath clear floor panels showing how the fossils appeared when discovered.

◀◀ Full-height window walls provide a strong visual connection between interior and exterior spaces.

SKY CONDITIONS

Cloudiness data was not available for the immediate vicinity, so this chart represents conditions in nearby Riverside.

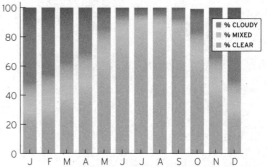

■ % CLOUDY
■ % MIXED
■ % CLEAR

TEMPERATURES & DEW POINTS

The area's intense summer heat makes the shading of surfaces a high priority.

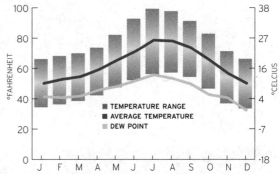

■ TEMPERATURE RANGE
■ AVERAGE TEMPERATURE
■ DEW POINT

HEATING/COOLING DEGREE DAYS

High temperatures coupled with solar gain make cooling the dominant comfort factor.

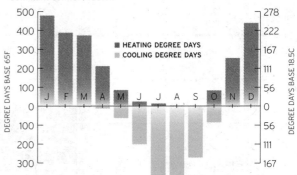

■ HEATING DEGREE DAYS
■ COOLING DEGREE DAYS

GROUND FLOOR

1 Museum lobby
2 Exhibition space
3 Cafe
4 Library
5 Office
6 Meeting room
7 Service yard
8 Outdoor exhibition
9 Loggia
10 Courtyard
11 Giftshop
12 Collections archive
13 Laboratory
14 Conference room
15 Administration lobby

SOURCES

METAL/GLASS CURTAINWALL: Centria

GLASS: Viracon VE1-2M Insulating Glass, Carlisle

COATINGS: Frazee, Majestic, Mirro, McClosky, USG, Flatwork–Davis, Shaw & Sons

CEILINGS: Chicago Metallic Planostile; Celotex Theatre Black

BATHROOM TILE: Quarry Tile Company Eco-Tile

CARPET TILES: InterfaceFlor Sabi

FLAT ROOF PV SYSTEM: Sharp PV Solar Electric Modules

INVERTERS AND MONITORING: Sunny Boy

MOUNTING SYSTEM: Unistrut

BIPV LOGGIA SYSTEM: Solar Electric Module WLM 375

DATA ACQUISITION: Heliotronics Aristotle

DUAL-FLUSH TOILETS: Caroma Walvi

URINALS: Sloan

LIGHTING CONTROLS: Leviton (Centura daylight dimming system)

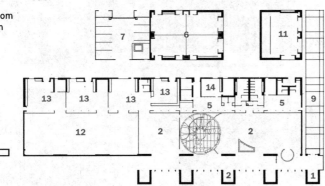

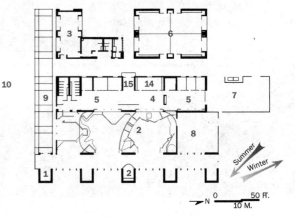

Let the Sun Shine In

A LIGHT-FILLED LIBRARY BECOMES A DYNAMIC MEETING PLACE FOR AN UNDERSERVED COMMUNITY

RUSSELL FORTMEYER

BRONX LIBRARY CENTER
BRONX, NY

THE SIDEWALK IN FRONT OF NEW YORK'S new Bronx Library Center is a good eight-feet wider than the rest of the block, all the better to accommodate the crowds of people who have flocked to the building since it opened. Elga Cace served as head librarian for nearly 25 years at the original building down the street. "It's time the Bronx had something like this," she says, referring to five floors now jam-packed with people flipping through books, checking e-mail, and engaging in boisterous conversation during a brisk fall Friday evening.

New York–based Dattner Architects intensified the community aspects of the 78,000-square-foot, $50 million building by centering its design around a four-story-high-performance glass curtain wall that fronts East Kingsbridge Road, just steps from the Bronx's busy Fordham Avenue shopping district. Reading areas are positioned along the curtain wall in a 16-foot-wide structural cantilever, resulting in a display of people that makes the building look more like a busy retail store than a traditional library.

"When we built this building, it was so open there was some question of how [it] would be received," says Dattner principal Daniel Heuberger, AIA. With that question answered handily by the mobs of people, Heuberger explains that achieving the openness without sacrificing energy efficiency was no accident.

The glass has a U-value of 0.39, relatively low and better than most dual-pane windows. A custom-integrated light shelf, with a built-in indirect fluorescent strip light, reflects the east-exposure morning light up to a vaulted ceiling that in turn forces daylight deep into the building's interior. Translucent nylon mesh mechanized shades, integrated into the wall system, manually operate at each floor's service desk. "When you have material assemblies that combine different building trades, you have to be very good about coordination," Heuberger says.

Ceiling-mounted photocells on every floor track lighting levels. They dim the compact fluorescent downlights along the glass and control bilevel switching at the fluo-

KEY PARAMETERS

LOCATION: Bronx, N.Y. (Hudson River watershed)

GROSS SQUARE FOOTAGE: 78,000 ft² (7,200 m²)

COST: $31 million

COMPLETED: January 2006

ANNUAL ENERGY USE (BASED ON SIMULATION): 49 kBtu/ft² (558 MJ/m²), 28% reduction from base case

ANNUAL CARBON FOOTPRINT (PREDICTED): 15 lbs. CO_2/ft²(72 kg CO_2/m²)

PROGRAM: Reading areas, stacks, computer workstations

BRONX LIBRARY TEAM

OWNER: The New York Public Library

ARCHITECT: Dattner Architects

LANDSCAPE: MKW & Associates

ENGINEERS: Robert Derector Associates (MEP); Severud Associates (structural); Langan Engineering (civil)

COMMISSIONING AGENT: Steven Winter Associates

SUSTAINABLE DESIGN CONSULTANT: Jonathan Rose & Companies

LIGHTING: Domingo Gonzalez Design

ACOUSTICAL, A/V, TELECOM: Shen Milsom & Wilke

COST ESTIMATOR: VJ Associates

OWNER'S REP: Walter Associates

CONSTRUCTION MANAGER: F.J. Sciame Construction

LEED SCORES
LEED-NC Version 2 Silver

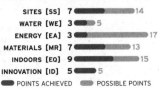

	Points Achieved	Possible Points
SITES [SS]	7	14
WATER [WE]	3	5
ENERGY [EA]	3	17
MATERIALS [MR]	7	13
INDOORS [EQ]	9	15
INNOVATION [ID]	5	5

● POINTS ACHIEVED   ● POSSIBLE POINTS

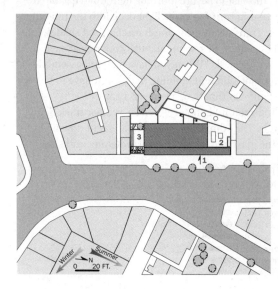

rescent direct/indirect lights that hang over book stacks. Seventy-five percent of the spaces meet LEED's minimum criteria for the ratio of daylight to illuminated light, and the effects of the daylighting scheme make the interior seem much brighter than the building's low lighting power density of 1.3 watts per-square-foot would suggest.

For the main reading room on the fourth floor, the architects wanted to "peel back" the ceiling up to the fifth-floor mezzanine to create a west-facing clerestory that would let late-afternoon sun flood the two floors. The swooping ceiling and roof element crown the building, establishing what Heuberger considers the library's claim to civic landmark status.

The library—the flagship for the Bronx's 34 branches—sits one block south of the home of Edgar Allan Poe and a block west of the original Fordham Library, which was a dark, technologically outdated building a third of the size of this one. A Con Edison utility building previously on the site was torn down; 80 percent of it was recycled, with the remainder sold to scrap dealers—an arrangement that paid for the cost of demolition. The property was excavated for a full basement that includes a 150-person-occupancy auditorium, conference rooms, and computer classrooms, as well as a large gathering area that brings daylight from an open stairwell to the ground floor. The artist Iñigo Manglano-Ovalle's "Portrait of a Young Reader," a work of colored-glass cylinders mounted on steel backing, sets off the staircase and links the two floors.

Elevators and a main stair are situated at the back and west side of the building (the stairs are clad in a frosted channel glass, lending privacy to the apartment buildings surrounding the library), and offices and support spaces are positioned along the west and north sides. In addition to the auditorium, community groups can reserve meeting rooms on each floor; eventually, public events will also take place on an outdoor terrace on the south side of the third floor. Minimal use of finishes in these areas contributes to the consistently open feeling of each floor. The architects wanted spaces and materials that were easily recognizable by patrons: The exposed wood for casework is a lightly finished Forest Stewardship Council (FSC)-certified maple; rough Minnesota red granite used on walls was water-treated to bring out its color; the elevator core is bright blue; and a maple veneer was applied to the main reading room's metal ceiling system. More than 55 percent of were sourced within 500 miles.

⌃ A wide staircase connects the ground floor to a large gathering space adjacent to the basement auditorium (top). Patrons find reading areas along the building's curtainwall bathed in sunlight (above). The smaller chairs and tables denote the second floor children's collection.

SITE PLAN

1 Entrance
2 Mechanical penthouse
3 Exterior roof garden

Winter Summer
N
0 20 FT.

SOURCES

ELEVATORS AND ESCALATORS: Thyssen Krupp

CEILINGS: Armstrong Ceilings Plus

CARPET: Mannington, Interface, Lees, Collins & Aikman

RESILIENT FLOORING: Forbo, Johnsonite

LIGHTING CONTROLS: ETC, Lutron

The roof "scoop" connects the fourth floor reading area curtain wall to the clerestory of the business resource center on the fifth floor. This design strategy allows the building to harness morning and afternoon daylight for the main reading room.

SKY CONDITIONS

New York skies are cloudy or partly cloudy much of the year, suggesting a wide-open approach to daylighting the interior.

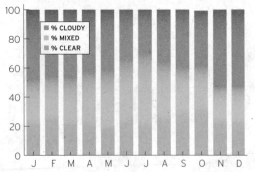

Legend:
- % CLOUDY
- % MIXED
- % CLEAR

(Bar chart, months J F M A M J J A S O N D, vertical axis 0–100)

TEMPERATURES & DEW POINTS

Temperatures range widely in New York, with only limited swing seasons with temperate conditions.

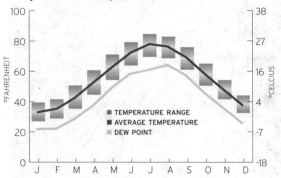

- TEMPERATURE RANGE
- AVERAGE TEMPERATURE
- DEW POINT

(Chart with °FAHRENHEIT axis 0–100, °CELCIUS axis -18 to 38, months J F M A M J J A S O N D)

HEATING/COOLING DEGREE DAYS

The number of cooling degree days is small compared to the heating degree days, but cooling represents a substantial load nevertheless.

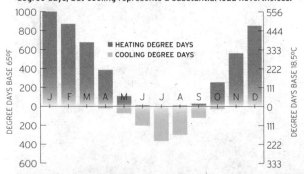

- HEATING DEGREE DAYS
- COOLING DEGREE DAYS

(Chart: DEGREE DAYS BASE 65°F axis; DEGREE DAYS BASE 18.5°C axis; months J F M A M J J A S O N D)

⌃
A controls system modulates the lighting in two zones: at the windows, dimmers shut off lights during peak sunshine hours; while stepped switching schemes shut off unnecessary lights over interior computer workstations and stacks.

«
The library's site in the middle of a bustling, diverse Bronx neighborhood ensures a steady stream of patrons throughout the day.

Mechanical rooms on the roof include air-cooled chillers and stacked air-handling units (AHUs), supplemented by boilers in a basement room. The decision to use air-cooled rather than water-cooled chillers was made after a series of conversations among the architects, client, and the project's mechanical engineer, David Peterman, of Robert Derector Associates. "With air-cooled chillers, there's no possibility of freezing, no loss of water, no plumes of moisture or chemicals, and less maintenance," Peterman says. And although a water-cooled system's cooling towers are typically more efficient, acoustical concerns, in addition to space requirements, ultimately led back to air-cooled chillers.

Other mechanical system features include high-efficiency filters and economizer modes for the AHUs, variable-speed drives for pumps, and compressors rated less than 15 horsepower so the library wouldn't have to hire more expensive licensed operators. Altogether the library's systems are designed to use 18.2 percent less energy measured against an ASHRAE Standard 90.1-1999 baseline.

Pat Konecky, a project manager with the New York Public Library's Office of Capital Planning and Construction, says the embrace of sustainability when

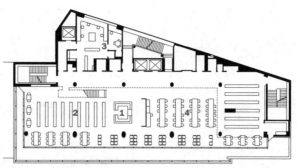

FOURTH LEVEL

1 Circulation desk
2 Collection
3 Staff area
4 Computer stations

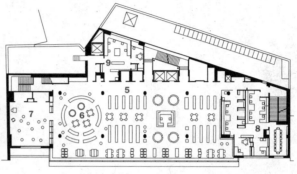

SECOND LEVEL

5 Children's collection
6 Homework area
7 Multipurpose room
8 Administration
9 Staff area

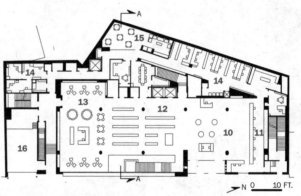

ENTRY LEVEL

10 Entrance lobby
11 Main circulation desk
12 Popular collection
13 Teen area
14 Staff area
15 Lounge
16 Loading bay

N 0 10 FT.

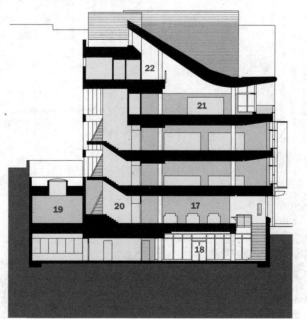

SECTION A-A

17 Lobby
18 Prefunction space
19 Staff area
20 Main stair
21 Reading room
22 Business skills center

0 10 FT.

design began in 2001 represented a departure for the library system. "Sustainability wasn't in the front of people's minds when this project started," she says. "Now, it is something we consider for every project we do." Konecky also says the library's design success has encouraged maintenance staff to do a better job of keeping the building up, even though they needed more extensive training to operate the building's complex mechanical control systems. And Steven Winter Associates, the library's sustainability consultants, have returned to provide additional commissioning during the first year of operation. "This has resulted in better-functioning equipment," Konecky says, noting that the additional commissioning has revealed only minor problems.

If the library's daylighting success overshadows the subtler sustainable characteristics of the project, no one seems to mind. The library system intended to build a resource for a community, not necessarily a high-performance sustainable building. Michael Alvarez, the new head librarian, has worked in nearly 25 libraries in his career, but none have been as inviting as the Bronx Library Center. "Most are very similar to one another," he says. "But the light-filled feel of this library was so radically different, I had to apply for this job." ≪

Ecological Literacy

A COPPER-INFUSED GLASS SKIN BLURS THIS LIBRARY'S BOUNDARIES

CASE STUDY
CENTRAL LIBRARY
DES MOINES, IOWA

NADAV MALIN

KEY PARAMETERS

LOCATION: Des Moines, Iowa (confluence of Raccoon and Des Moines rivers)

GROSS SQUARE FOOTAGE: 145,500 ft² (13,520 m²)

COST: $30 million

COMPLETED: December 2005

ANNUAL PURCHASED ENERGY USE (EXTRAPOLATED FROM TEN MONTHS' UTILITY BILLS): 88.6 kBtu/ft² (1,000 MJ/m²)

ANNUAL CARBON FOOTPRINT: (PREDICTED) : 40 lbs. CO_2/ft² (197 kg CO_2/m²)

PROGRAM: Reading areas, stacks, offices

TEAM

OWNER: Des Moines Public Library

ARCHITECT: David Chipperfield Architects

ASSOCIATE ARCHITECT: Herbert Lewis Kruse Blunck Architecture

FAÇADE CONSULTANT: W.J. Higgins & Associates

ENGINEERS: Arup (MEP), KJWW Engineering Consultants (MEP), Jane Wernick Associates and Shuck-Britson (structural)

COMMISSIONING AGENT: KJWW

GENERAL CONTRACTOR: The Weitz Company

WHEN YOU'RE INSIDE THE BUILDING, IT FEELS LIKE YOU'RE sitting in the park," says Paul Mankins, who was the Partner in Charge for Herbert Lewis Kruse Blunck Architecture (HLKB) on the new Des Moines Public Library. Mankins, who is now a partner at Substance Architecture, was referring to the building's glass skin, which he considers the most remarkable element of the deceptively simple-looking library. It's an ethereal skin designed to blur the boundary between indoors and out. Achieving that effect while providing a respectable thermal envelope forced the designers to collaborate with the glazing manufacturer to incorporate shading—in the form of a fine expanded copper mesh—into the triple-glazed, 4-by-14-foot panels, and then to install those panels without exterior mullions. Mankins admits that his team hadn't fully anticipated the visual effect of the glazing: "At sunset you see the building go from opaque copper to transparency. It dematerializes."

Having prevailed in a competitive selection process, London-based David Chipperfield Architects offered four distinct concepts from which the city could choose. Chipperfield's presentation to the selection committee was carried on local public access television (and repeated several times on the content-hungry station), and the public weighed in. The library selected the scheme that had also proven most popular with the public—a curvilinear two-story form with tentacles stretching into a surrounding park. The slender fingers of that design "had to be bulked up to con-

« The two-story library stretches across one end of Des Moines' five-block-long Western Gateway Park.

SKY CONDITIONS

Clouds soften Des Moines' skies at least half the time, except in the late summer and early fall.

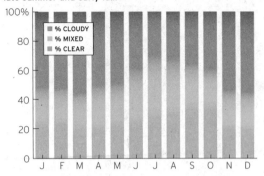

Legend:
- % CLOUDY
- % MIXED
- % CLEAR

(Y-axis: 100%, 80, 60, 40, 20, 0; X-axis: J F M A M J J A S O N D)

TEMPERATURES & DEW POINTS

Outdoor temperatures tend to extremes in Des Moines, with both the winter chill and summer heat exacerbated by humidity.

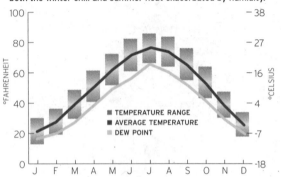

Legend:
- TEMPERATURE RANGE
- AVERAGE TEMPERATURE
- DEW POINT

(Left Y-axis °FAHRENHEIT: 100, 80, 60, 40, 20, 0; Right Y-axis °CELSIUS: 38, 27, 16, 4, -7, -18; X-axis: J F M A M J J A S O N D)

HEATING/COOLING DEGREE DAYS

Consistent with the temperature trends, winter heating loads and summer cooling demands are substantial.

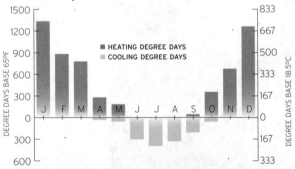

Legend:
- HEATING DEGREE DAYS
- COOLING DEGREE DAYS

(Left Y-axis DEGREE DAYS BASE 65°F: 1500, 1200, 900, 600, 300, 0, 300, 600; Right Y-axis DEGREE DAYS BASE 18.5°C: 833, 667, 500, 333, 167, 0, 167, 333; X-axis: J F M A M J J A S O N D)

tain the collection," says Mankins, but the general layout remained intact.

"The unusual shape allows for a lot of privacy for our customers," says acting director Dorothy Kelly. Compared with the one large reading room of the previous library, the new space offers visitors a much better chance of finding a quiet corner, even though visitation has increased threefold since the building opened in April of 2006. "One of the things that our customers, as well as our staff, particularly like is that there is so much light coming in," says Kelly.

After Chipperfield and its consultant, Arup, created the concepts, they collaborated with associate architect HLKB, engineer of record KJWW and the other consultants, to develop and document the design. Realizing Chipperfield's minimalist aesthetic required meticulous attention to details. "A lot of time was spent on alignments [of planes and surfaces], to make sure that those made sense," says Jeff Wagner of HLKB.

The green roof was not in the original design, but was added at the request of a neighboring office overlooking the building. That amenity will lend aesthetic value to others over time, according to Mankins, who notes: "The library is in the middle of what is going to be a densely developed part of Des Moines." Beyond the aesthetics, however, the green roof provides a valuable stormwater management function by modulating runoff. "This part of Des Moines has combined sewer and stormwater overflow," notes Mankins—when the stormwater system's capacity is exceeded, raw sewage is released into the river.

The choice to include a green roof affected the facades as well, according to Brett Mendenhall, who was a project architect for HLKB, and is now with OPN Architects. Because of the open interior plan with few full-height walls, there were limited locations for roof drains. Fewer drains would require longer and higher slopes on the roof. This meant the parapets would have needed to be raised significantly to accommodate a conventional sloped roof drainage system. "We knew that there were manufacturers who would guarantee the waterproofing on a dead flat roof," says Mendenhall, so the designers chose to eliminate the pitch entirely. The green roof was bid as an alternate, however, so when they made the decision to use a flat roof they were taking a risk. If the green roof were not implemented, "we would have had a big flat bathtub," notes Mendenhall. Fortunately, the green roof survived, thanks to a favorable bidding environment, and to the fact that lowering the parapet height reduced the size of the facades, saving enough money on the glazing panels to offset half the cost of the green roof.

Extending the minimalist aesthetic to the interior meant using a raised floor system with underfloor air distribution to limit clutter at the ceiling level, or what Scott Bowman of KJWW calls "shifting money from ceiling to floor." The system delivers air at constant volume in the central core, while over 200 fan-coil units at the perimeter provide heating or cooling as needed, an approach

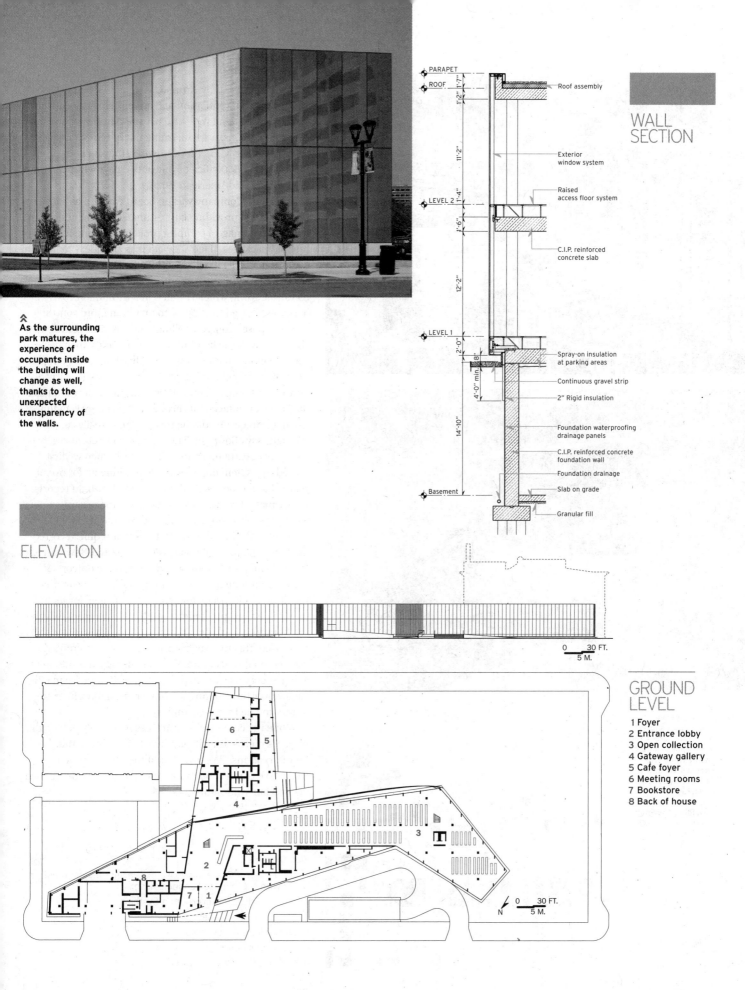

As the surrounding park matures, the experience of occupants inside the building will change as well, thanks to the unexpected transparency of the walls.

WALL SECTION

PARAPET
ROOF
1'-7"
1'-2"
Roof assembly

11'-2"
Exterior window system

LEVEL 2
1'-4"
Raised access floor system

1'-6"
C.I.P. reinforced concrete slab

12'-2"

LEVEL 1
2'-0"
8"
Spray-on insulation at parking areas

4'-0" min.
Continuous gravel strip

2" Rigid insulation

14'-10"
Foundation waterproofing drainage panels

C.I.P. reinforced concrete foundation wall

Foundation drainage

Basement
Slab on grade

Granular fill

ELEVATION

0   30 FT.
5 M.

GROUND LEVEL

1 Foyer
2 Entrance lobby
3 Open collection
4 Gateway gallery
5 Cafe foyer
6 Meeting rooms
7 Bookstore
8 Back of house

0   30 FT.
N   5 M.

The view from inside (top) shows how transparent the walls appear during the daytime. The exposed structure required exacting construction practices, while providing thermal mass that stabilizes temperatures and enhances comfort.

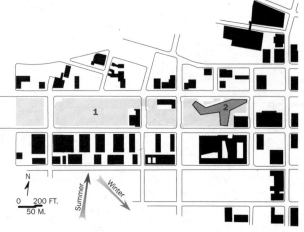

SITE PLAN

1 Western Gateway Park
2 Central Library

N

0   200 FT.
50 M.

Summer   Winter

that was suggested by Arup's U.K. team. Demand-based controls for the ventilation and an enthalpy wheel, which exchanges heat and moisture between incoming and outgoing air streams, were instrumental in reining in energy costs. Together with other measures, they helped win the project a $42,650 efficiency incentive payment from the local utility, MidAmerican Energy.

The Weitz Company was on board early on as construction manager, providing cost estimates during design, and helping to manage a complex process involving 22 separate bid packages. "Had they bid it as a single $25 million project, they only would have had a couple of bidders," says Mankins. Instead, many smaller contractors competed for the work, which resulted in a $1 milllion cost savings and allowed the library to maintain more control.

The exposed concrete ceilings and columns contribute significantly to the building's thermal mass, which helps to stabilize indoor temperatures. "It's hard to move [temperatures in] that building," notes Bowman. Having learned how the building responds thermally, the facility managers now take advantage of its benefits: "No energy is put in when it is unoccupied. The thermal mass really carries it through," says Bowman. The exposed concrete also generated a lot of extra moisture as it cured. "When we first turned the system on, we were dehumidifying for over a year—that has now settled down a bit," Bowman reports.

Bowman's firm was also responsible for commissioning, which, given the project's size and complexity, went smoothly. "We found only 50 items that required correction," he said. Most of them were minor and Bowman felt that planning to commission improved the outcomes.

Based on nine months of utility bills, it appears that the library's electricity and gas usage savings for its first year will exceed predictions by about 23 percent. Given the library's popularity and the extended hours that it has been used, this rate suggests that the energy simulation was reasonably accurate. With further attention to managing energy use, the building may well improve in efficiency over time, taking it from a reasonably efficient library to one that is exemplary.

While the community continues to actively debate the project's merits as a city landmark, the library staff has no reservations. "We love it, and the public loves it," says Kelly. "We've got tremendous circulation—it's being used beyond our wildest dreams." ◀◀

SOURCES

METAL/GLASS CURTAINWALL: Custom made by Architectural Wall Systems

GLASS: Facade - Okatech by Okalux Glazing (insulating glass panel with triple-glazed copper mesh)

ROOFING: Carlisle Waterproofing Systems

GREEN ROOF: Roofscapes

FLOORING: Tate Access Floors

CARPET: Lees Commercial Carpet

ELEVATORS/ESCALATORS: KONE

Uniform sizing of the glazed panel made mass-production possible and has led to significant cost savings.

Chapter 3

EDUCATION

THE TREND IN THE DESIGN OF EDUCATIONAL FACILITIES IS TO USE A BUILDING'S features as teaching tools. Many of the five case studies discussed in this chapter are part of this growing practice. For instance, Camp Arroyo, a multi-building complex in Livermore, California, provides weeklong environmental education programs for students from area schools and also serves as the setting for a summer camp for children with life-threatening illnesses. The buildings on the 138-acre campus deploy a range of structural systems: stabilized earth, steel frame with straw-bale in-fill, and wood framing. Each of these techniques offers an opportunity for staff to teach campers about climate-responsive solutions to differing programmatic requirements. The expanded campus of Sidwell Friends Middle School in Washington, D.C., is organized around its own on-site wastewater management facility. In the school's central courtyard is a constructed wetland that contains more than 80 species of native plants, filters, and settling tanks. This wetland and the treatment process it performs serve almost as a "new faculty member," suggests the project's landscape architect.

The pedagogical mission of the other projects presented in this chapter may not be quite as pronounced, but the buildings are nevertheless green to their cores. Students and faculty at the University of Oregon's Lundquist College of Business wanted their new 137,000-square-foot academic building on the Eugene campus to be green since they had identified sustainability as a key issue for companies of the future. To that end, the design team created a facility that utilizes different ventilation strategies in different spaces and one that is highly dependent on daylight for illumination. At the University of Washington in Seattle, the Center for Urban Horticulture's faculty and staff decided that the new Merrill Hall, which would replace a building destroyed by arson in 2001, should reflect the values of the programs housed within. The resulting building is cooled primarily by passive ventilation, includes a 9.6-kW photovoltaic array, and is the first facility on campus with a whole-site water system, linked to a two-decade-old bioswale.

The buildings included in this chapter do more than convey an image or help educators illustrate the relationship between the built and natural world. Projects like the Thomas L. Wells Public School, a 670-student kindergarten-through-eighth-grade school near Toronto, show that high-performance facilities are energy-efficient, comfortable, and filled with daylight. Above all else, the Wells School and the other buildings examined here provide settings that support both learning and teaching. «

CASE STUDY
CAMP ARROYO
LIVERMORE,
CALIFORNIA

NADAV MALIN

Eco-design Laboratory

RESPONDING TO THE HOT, DRY CLIMATE of the hills of California's East Bay, Siegel & Strain Architects selected three structural systems for the design of an educational camp. Each structural option offered a climate-responsive solution to the differing programmatic requirements, while enhancing the camp's value as an educational resource. The program included a dining hall to seat 200 campers, served by a commercial kitchen; two bathhouses adjacent to a swimming pool; and cabins to house 144 campers and staff. Additional space needs, including longer-term staff housing, were identified but excluded from the project due to budgetary constraints. These needs have since been met with conventional buildings.

Nothing about the Camp Arroyo project was simple,

beginning with the client group, which consisted of two separate organizations that planned to share the facility. Much of the focus on sustainability came from staff of the East Bay Regional Park District, which codeveloped the camp and intended to use it during the school year to run weeklong environmental education programs for children from area schools. The other client, the Taylor Family Foundation, was amenable to making it a green project but had other priorities as well. The foundation runs summer programs for children with life-threatening illnesses, so amenities such as the swimming pool, lawns, and air-conditioning were deemed critical.

Principal-in-Charge Larry Strain describes Camp Arroyo as a "breakthrough project for the firm." Siegel & Strain Architects was selected to design the project in spite of the fact that everything the firm had done previously was much smaller. "None of the four firms on the short list had proficiency with a project of this size," notes Strain. "They were clearly going for green experience and were willing to give up a background with projects of this scale to get that expertise."

The 138-acre location once housed a tuberculosis sana-

A MULTIBUILDING CAMP
FOR ENVIRONMENTAL
EDUCATION PROGRAMS SERVES
AS A WORKING LAB FOR TEACHING
GREEN-DESIGN PRINCIPLES

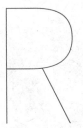

«
Even though air flows freely through the bathhouse, its thick earthen walls provide comfort on hot summer afternoons.

KEY PARAMETERS
LOCATION: Livermore Calif. (Arroyo Del Valle watershed)
GROSS SQUARE FOOTAGE: 20,000 ft² (1,860 m²)
COMPLETED: July, 2001
COST: $6.5 million
ANNUAL ENERGY USE (BASED ON UTILITY BILLS FOR THE DINING HALL AND CABINS): 88 kBtu/ft² (998 MJ/m²)
ANNUAL CARBON FOOTPRINT (BASED ON UTILITY BILLS): 18 lbs CO₂/ft² (86 kg CO₂/m²)
PROGRAM: Cabins, dining hall, kitchen, swimming pool with bathhouses

ARROYO TEAM
OWNER: East Bay Regional Park District
ARCHITECT: Siegel & Strain Architects
ENGINEER: Bruce King (structural); Davis Energy Group (MEP and energy)
LIGHTING CONSULTANT: After Image and Space
GENERAL CONTRACTOR: Antrim Construction
STRAWBALE SUBCONTRACTOR: Benchmark Development
STABILIZED EARTH SUBCONTRACTOR: Rammed Earth Works

EDUCATION **37**

SOURCES

SIDING: Hardieplank by James Hardie Building Products

WINDOWS: Dining Hall - custom FSC-certified mahogany and glass; Cabins - Metal-clad wood windows by Caradco

DOORS: Dining Hall - custom FSC-certified mahogany and glass; cabins - wood flush doors with FSC-certified cores

ROOFING: Bathhouse and cabins - Galvalume; Dining Hall - Pre-weathered Galvalume

PAINTS AND STAINS: Interiors - Sherwin Williams HealthSpec

PANELING: Cabins - wheat straw board manufactured by Primeboard

SPECIAL SURFACING: Toilet partitions of all buildings, recycled plastic, Santana

FLOOR AND WALL TILE: Terra Green Ceramics

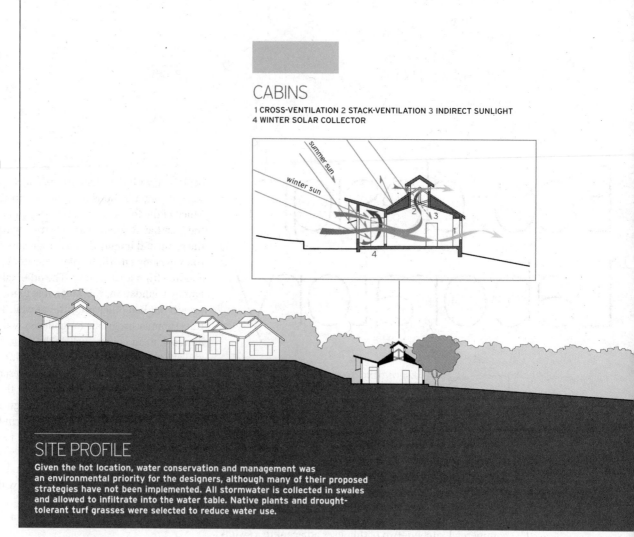

CABINS

1 CROSS-VENTILATION 2 STACK-VENTILATION 3 INDIRECT SUNLIGHT 4 WINTER SOLAR COLLECTOR

summer sun
winter sun

SITE PROFILE

Given the hot location, water conservation and management was an environmental priority for the designers, although many of their proposed strategies have not been implemented. All stormwater is collected in swales and allowed to infiltrate into the water table. Native plants and drought-tolerant turf grasses were selected to reduce water use.

torium. New buildings were sited exclusively in areas that had already been disturbed by prior construction.

Siegel & Strain focused first on building orientation and form for energy efficiency. The team subsequently selected different structural systems for each of the three types of buildings: stabilized earth for the bathhouses, steel frame with straw-bale infill walls for the dining hall, and efficient wood framing for the cabins. Light-colored corrugated metal roofing with large overhangs on all the buildings helps to unify the project visually.

The bathhouse walls are made of cement-reinforced earth, colored to match the soil, connecting the building with the site. The potentially low-embodied energy of this wall system was compromised: First, the soil proved to have very high clay content, so sand was trucked in to augment the mix and give it an appropriate consistency for construction; second and more important, in this high-risk earthquake zone, the walls required a high percentage of cement and had to be reinforced with steel, so their ecological profile is similar to that of standard concrete walls.

For the dining hall, thermal performance to keep out the summer heat was a high priority, and the client had an interest in straw-bale construction. Siegel & Strain turned to structural engineer Bruce King, author of *Buildings of Earth and Straw* (Ecological Design Press, 1997) for the necessary expertise. The bale walls were coated on either side with 1 1/2 inches of gunite and a layer of plaster, giving them significant thermal mass in addition to the thermal resistance of the bales.

The team felt the thickness of straw-bale walls would feel out of proportion in the duplex cabins, according to project architect Nancy Malone. Instead, they specified 2-by-6 studs and cellulose insulation, with advanced framing to avoid excessive wood use. All of the framing lumber and sheathing were certified according to the standards of the Forest Stewardship Council (FSC).

The designers also developed a sophisticated on-site wastewater treatment system using constructed wetlands that would have been used to irrigate a garden in an old walnut orchard. In spite of the designers' success in

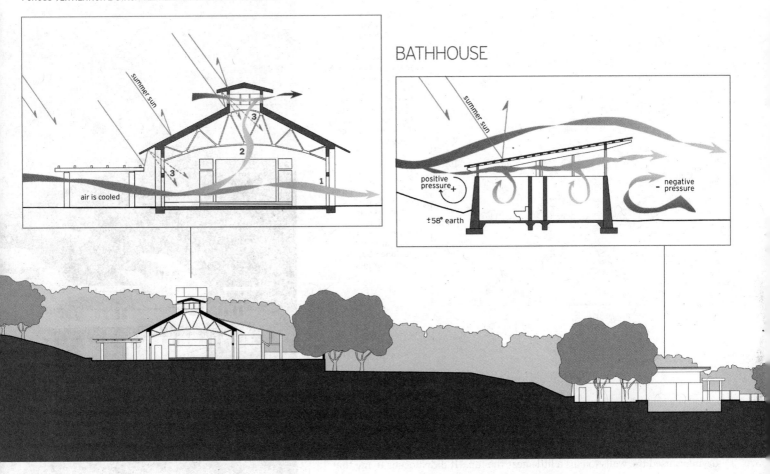

DINING HALL

1 CROSS-VENTILATION 2 STACK-VENTILATION 3 INDIRECT SUNLIGHT

summer sun

3

2

3

1

air is cooled

BATHHOUSE

summer sun

positive pressure +

negative pressure

±58° earth

obtaining the permits for that system, the Park District ultimately opted for a more conventional leach field for wastewater due to maintenance concerns. "The conventional system probably requires just as much maintenance," says Malone, "but it was maintenance that they understood, as opposed to something new."

An ambitious permaculture-based landscape plan was also scoped out, but was dropped during design as a cost-control measure. Camp Arroyo still intends to develop its landscape along those lines, according to its program director, Kathy Swartz, and in the meantime maintains an organic vegetable garden.

Quality control was a challenge through much of the project, notes Malone. There was a problem with

concrete slabs curling at the edges as they cured, for example. While that could be blamed on the fact that coal flyash replaced 50 percent of the cement in the concrete, Malone points out that the contractors didn't take standard precautions to ensure an even cure, in spite of the dry heat.

In some cases even a strong commitment from the contractor didn't go far enough to guarantee compliance with the ecological program. When it came to building the cabins, the framing contractor embraced the optimum value engineering approach to minimize wood use. He happened to be absent when the lumber was delivered, however, and the framing crew dove right in. By the time he returned, two of the cabins were largely framed out using standard framing.

Siegel & Strain focused first on building orientation and form to achieve energy efficiency.

SITE PLAN

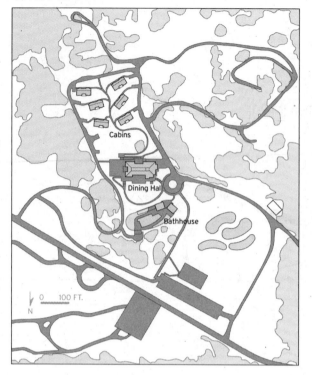

Cabins

Dining Hall

Bathhouse

0   100 FT.
N

» Overhangs were carefully sized to keep summer sun off the south-facing windows. Inside, placards explain how to manually control the windows and shutters to maintain comfort during different weather conditions.

« Kidney-shaped ponds at the north end were part of a construction wetland waste-water treatment system that was designed but not implemented.

While structurally sound, the earthen bathhouse walls developed a problem shortly after construction. Steel reinforcing was used near the interior face but not at the exterior. As a result, shrinkage of the clay and silt during drying caused the walls to curl a little near the top. "It didn't occur to me that the wall might shrink enough to curl the wall back," says King. "If I were doing those walls again, I would add rebar on the outside."

Even with the challenges, the completed project has been a huge success, winning recognition from the AIA as a "Top Ten Green Project" in 2002. That same year, the YMCA of the East Bay took over management of Camp Arroyo, although the East Bay Regional Park District and the Taylor Family Foundation continue to run their programs at the facility. "The buildings are such a teaching tool for us," says Swartz. She points to the range of structural systems, in particular, as a valuable tool for teaching campers about sustainable design.

Performance issues continue to plague some of the mechanical systems in the dining hall and cabins. Fortunately, the real success of the design is that those systems are hardly needed, even on the hottest summer days. Dr. Gail Brager's Sustainable Design for Hot Climates class from the University of California at Berkeley's College of Environmental Design spent 19 days studying the buildings in 2002. With support from the architects and the Pacific Energy Center, they tracked temperatures indoors and out, on various surfaces and in the middle of the spaces. Their results show that even though the evaporative cooler in the dining hall wasn't functioning properly, conditions remained comfortable indoors through several very hot afternoons. «

HEATING/COOLING DEGREE DAYS

The camp is little used during the peak heating months of december and january, and used intensively during the peak cooling months.

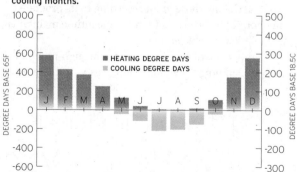

■ HEATING DEGREE DAYS
■ COOLING DEGREE DAYS

J F M A M J J A S O N D

The strawbale walls in the dining hall were constructed by Rick Green of Willows, California, a third-generation rice farmer and supplier of bales for many building projects in the state. The walls were then covered with stabilized earth and plaster.

Earth construction guru David Easton of Napa, Calif., built the bathhouse walls using a method he pioneered, in which he shoots the material out of a hose, gunite-style, against a one-sided form. This method, also called pneumatically installed stabilized earth (PISE), offers many of the advantages of rammed earth with lower labor cost.

TEMPERATURES & DEW POINTS

The normal high, average, and low temperatures shown for each month don't reflect the extremes that are often experienced.

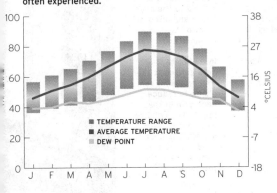

- TEMPERATURE RANGE
- AVERAGE TEMPERATURE
- DEW POINT

SKY CONDITIONS

Cloudiness data was not available for livermore, so this chart is for stockton, Calif., which experiences more winter cloudiness.

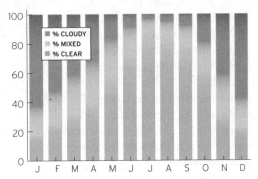

- % CLOUDY
- % MIXED
- % CLEAR

Creative Accounting

CASE STUDY
LILLIS BUSINESS COMPLEX, UNIVERSITY OF OREGON, EUGENE, OREGON

NADAV MALIN

THE AMBITIOUS USE OF DAYLIGHTING AND PHOTOVOLTAICS IN THIS BUILDING CREATES A BUSINESS CASE FOR GREEN DESIGN

THE LILLIS BUSINESS COMPLEX WAS FIRST CONCEIVED as a small addition to the University of Oregon's Lundquist College of Business to relieve some space constraints and provide new classrooms. During the programming phase, however, the notion of tearing down a two-story brick connector and replacing it with something larger was proposed. That connector, which obstructed a main circulation route, was "one of the more reviled buildings on campus," says Fred Tepfer, of the university's planning office. The new complex, with its four-story atrium, provides much more space and amenities than were originally envisioned and unclogs the circulation route at the same time.

As the scope of the project grew, so did the project team's green aspirations. Faculty and students from the College of Business had identified sustainability as a key business strategy for companies of the future. "They came to us asking for a building that showed them how to think in a fresh way about those business decisions," reports Kent Duffy, AIA, design principal for the project for SRG Partnership. While SRG had implemented various green measures on its projects, the company "hadn't had a chance to put them together in one building," according to Duffy. He jumped at the opportunity, and describes the process that ensued as "a career-transforming experience."

A multidisciplinary design team worked collaboratively on the Lillis project from the beginning. Many of the green goals were both proposed and developed by G. Z. (Charlie) Brown, FAIA, a professor in the university's Department of Architecture and the director of the Energy Studies in Buildings Laboratory. Oregon requires the engagement of a construction manager during design for

⌃ When no UL-approved mullion system could be found in which to run the wires, the electrical inspector agreed to approve the curtainwall installation as a prototype.

≫ Photovoltaics are affixed to the skylights and integrated into the atrium's south-facing curtainwall in a pattern that gets denser at the top, providing some shade to reduce heat gain in the space. The building has a total of 45 kW of PVs, most of which are in conventional panels on the roof.

RICK KEATING

KEY PARAMETERS

LOCATION: Eugene, Oregon (Willamette River watershed)

GROSS SQUARE FOOTAGE: 137,000 ft^2 (12,700 m^2)

COMPLETED: October 2003

COST: $41 million

ANNUAL ENERGY USE (BASED ON SIMULATION): 62 kBtu/ft^2 (705 MJ/m^2)–39% reduction from base case.

ANNUAL CARBON FOOTPRINT (PREDICTED): 10 lbs. CO$_2$/ft^2 (50 kg CO$_2$/m^2).

PROGRAM: A four-story addition connecting three preexisting buildings, consisting of an atrium/thoroughfare, a cafe, public meeting rooms, classrooms, and offices.

LILLIS TEAM

OWNER: University of Oregon

ARCHITECT AND INTERIOR DESIGNER: SRG Partnership

COMMISSIONING AGENT: Solarc Architecture & Engineering

ENGINEER: Degenkolb Engineers (structural); Balzhiser & Hubbard Engineers (MEP)

LANDSCAPE ARCHITECT: Cameron McCarthy Gilbert Scheibe

ENERGY/COMMISSIONING: Solarc Architecture & Engineering

LIGHTING: Benya Lighting Design

ACOUSTICAL: Altermatt Associates

PHOTOVOLTAIC SYSTEMS: Solar Design Associates

GENERAL CONTRACTOR: Lease Crutcher Lewis

LEED RESULTS
LEED-NC Version 2 Silver

SITES [SS]	8	14
WATER [WE]	1	5
ENERGY [EA]	9	17
MATERIALS [MR]	6	13
INDOORS [EQ]	5	15
INNOVATION [ID]	4	5

POINTS ACHIEVED   POSSIBLE POINTS

⌃
The new courtyard at Lillis, a popular hangout, features a specimen Yellow Buckeye that was a gift from the governor of Ohio and Ohio State University in fulfillment of a bet on the 1958 Rose Bowl Game.

large state-funded projects. This model "works well if you have a good team," says Matt Pearson, project manager for the general contractor Lease Crutcher Lewis. "It's rare that you have a team work as well as that team did."

Although tightly constrained on all sides, the site is elongated from east to west, making it well oriented for daylighting. Initially, the faculty was skeptical about the potential for bringing daylight into the classrooms due to concerns about contrast on the projection surfaces at the front of the rooms. Under Brown's direction, students created a computerized daylighting simulation showing the distribution of light levels in a classroom. In addition, a classroom was mocked up in the existing building prior to its demolition to give users a chance to experience the proposed space. After the study revealed that flipping the orientation of the classrooms would allow for plenty of light in the seating area, while keeping the projection surfaces dark, the College of Business faculty bought the idea. Upon seeing this solution for the business school classrooms on the second floor, the university chose a similar layout for its general-use classrooms on the first floor as well.

Brown's team also worked with lighting designer James Benya and lighting controls supplier Lutron Electronics to develop the controls strategy. Lillis repre-

sents the first time that Lutron provided integrated controls for lighting and shading devices, according to Duffy, and the results set a new direction for the company. The shades expand upward from the bottom of the windows, keeping the tops of the windows exposed for optimum daylighting. Flipping a light switch in the classroom opens the shades and turns on the lights, but the lights are dimmed to 10 percent of their full brightness unless additional illumination is needed.

Eugene's relatively mild climate makes it ideal for natural ventilation and night flushing. The building uses different ventilation strategies in different locations, including 100 percent natural ventilation (no mechanical cooling or ventilation air) in the atrium and faculty offices on the north, hybrid natural and mechanical ventilation and cooling in the classrooms, and 100 percent mechanical ventilation and cooling in the faculty offices on the south.

Extensive computer modeling revealed that the thermal mass needed for the night flushing strategy could be met with a steel structure by adding thin slabs in key locations. Without this modeling, a concrete structure might have been selected, adding more than a million dollars to the cost. Mass was added to selected indoor surfaces, and special plenums were created

ATRIUM VENTILATION DIAGRAM

1 Atrium
2 Balconies
3 Classrooms

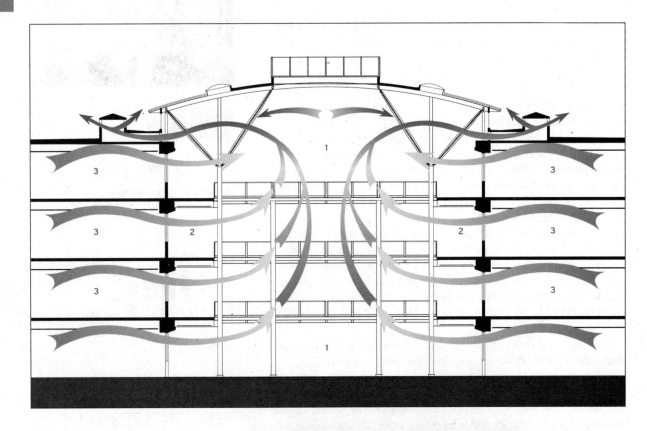

» **Interior and exterior lightshelves reflect daylight deep into the classrooms while protecting the zone near the perimeter from glare.**

SITE PLAN

1 Existing Lillis 2 Existing buildings 3 Rotunda 4 Atrium
5 Auditorium 6 Classrooms/offices

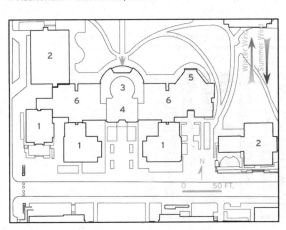

SOURCES

METAL/GLASS CURTAINWALL: Vistawall CW-600 (with PV)

ROOFING: Stevens Hi-Tuff EP Fleece adhered TPO

INTERIOR AMBIENT LIGHTING: Smedmarks Minisize T5; Zumbtobel/Staff Claris; Finelite Series 8; Translite Systems Liana; Leucos Modulo

CONTROLS (INTEGRATED LIGHT & SHADE): Lutron Grafik 6000

PHOTOVOLTAICS: Custom glass-integrated PVs: Saint-Gobain (curtainwall & skylights)

FLAT ROOFTOP POLYCRYSTALLINE PV ARRAYS: Sharp

PEEL AND STICK PV MEMBRANE: UniSolar

MOTORIZED WINDOW SHADES: Lutron Sivoia QED

to maximize the contact between the ventilation and the surfaces of the thermal storage mass.

Financing the photovoltaic (PV) systems was a challenge that was overcome only when a representative of the local utility convinced the state energy office to allow the university to transfer tax credits to an outside donor. Ultimately, the building-integrated PVs in the curtainwall make up just 13 percent of the building's total of 45 kW of PVs. In reality they deliver even less, because a big yellow buckeye tree largely shades half the wall. (PV cells in a panel deliver only as much electricity as the least productive cell in each row, so shading a panel reduces its output more significantly than the amount of shading might suggest.) Most of the solar electricity at Lillis is generated by a large array of conventional panels mounted flat on the rooftop; the four systems together supply about 10 percent of the building's predicted electricity demand.

Lease Crutcher Lewis managed the construction waste by sorting for valuable materials, such as metals and cardboard, on-site, and commingling the remainder, which was then sorted at a dedicated facility off-site. "Eugene is unique—one of the easiest places to recycle waste," says Pearson. The preexisting building that was demolished to make room for Lillis was largely ground into rubble, which was used as fill for other projects in town.

While most users of Lillis are thrilled with the building, a few situations have created significant challenges for the facility managers. Tepfer is frustrated that the commissioning process took a long time to complete, leading to complaints from some occupants about problems that should have been resolved prior to occupancy. The natural ventilation and night flushing strategies have been especially challenging for a commissioning process that typically focuses on mechanical equipment. Students who investigated the building under Professor Alison Kowk, using simple, homegrown tools, "have provided more useful information than the commissioning," complains Tepfer.

To some extent, complaints about comfort are exacerbated by a design decision to provide offices on the south side of the corridor with mechanical cooling and fixed windows, while offices on the north got operable windows but no air-conditioning—a situation that Tepfer describes as "a Faustian deal."

Anticipating that the stack effect in the atrium may not consistently provide enough of a pressure differential to drive the natural ventilation process, the design team identified smoke evacuation fans in the atrium roof as a means of enhancing the airflow. The fans were outfitted with variable-speed drives so that they could be operated at low speed to improve the ventilation without a large energy penalty. Unfortunately, the controls on those drives were not configured in a way that activates the fans appropriately; once the control strategy for those fans is refined, the comfort on the upper floors should improve. Similarly, the fourth-floor auditorium is being retrofitted with ceiling fans, according to Duffy, to alleviate summertime discomfort: "We thought that the stack effect was going to be so successful that we wouldn't need those fans."

Mostly, though, the university's experience with the building has been overwhelmingly positive. "People really love the building," says Tepfer. "Good daylighting makes happy people." G. Z. Brown agrees: "The big atrium space in the middle is filled with people all the time," and the toplit lecture hall on the top floor is "one of the more sought-after lecture halls on campus." «

« The atrium space includes a popular cafe.

HEATING/COOLING DEGREE DAYS

Degree days are calculated based on daily average temperatures, so cool nights can mask uncomfortable daytime high temperatures.

HEATING DEGREE DAYS
COOLING DEGREE DAYS

TEMPERATURES & DEW POINTS

The wide range between the normal high and low temperatures in summer illustrates the potential for night-flushing as a cooling strategy.

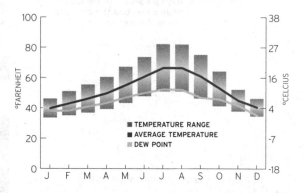

TEMPERATURE RANGE
AVERAGE TEMPERATURE
DEW POINT

SKY CONDITIONS

Cloudiness data was unavailable for Eugene, Oregon, so the information shown here is for Portland, Oregon.

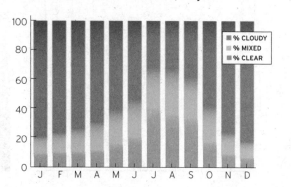

% CLOUDY
% MIXED
% CLEAR

«
The design provided for a combination of fans and passive airflow to circulate air from classrooms and offices into the atrium, where the natural stack effect is supposed to drive the air out through vents in the roofs. Using soap bubbles, students discovered that the air doesn't always cooperate.

The new complex has become a hub for campus activities and a major thoroughfare, replacing a smaller builidng that impeded the flow.
⌄

"The big atrium space in the middle is filled with people all the time."

SIDWELL FRIENDS MIDDLE SCHOOL, WASHINGTON, D.C.

NADAV MALIN

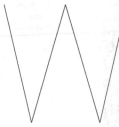HILE STUDYING AERIAL photographs of the hilltop campus of Sidwell Friends Middle School, the project team recognized the campus sits atop two watersheds, both of significant ecological value. This insight led to an integrated approach to water management as the centerpiece of a comprehensive appeal to environmental stewardship that emerged through encounters with architect William McDonough, FAIA, and educator David Orr. "We started out designing a building, which turned into a green building, and that green building ended up transforming the whole school, culturally and operationally," says Mike Saxenian, assistant head of the school and its chief financial officer.

Sidwell Friends School is split between two campuses. Children in pre-kindergarten through fourth grade attend the lower school on the Bethesda, Maryland, campus. Older students go to the Washington, D.C., campus four miles to the south, which houses the middle and upper schools. A comprehensive master-planning process for both campuses, led by Philadelphia-based KieranTimberlake Associates (KTA), determined that updating and expanding the 55-year-old middle school was the first priority. Following presentations from several short-listed firms, the school hired KTA to design the project.

To create the new middle school, the design team renovated the existing 33,000 square-foot building and expanded it with a 39,000-square-foot addition. The old and new wings meet to form a U-shaped courtyard. The primary entrance leads through the courtyard into a spacious lobby, which, together with administrative offices, connects the old and new parts of the building.

LEED SCORES
LEED-NC Version 2 Platinum

	Points Achieved	Possible Points
SITES [SS]	11	14
WATER [WE]	5	5
ENERGY [EA]	13	17
MATERIALS [MR]	8	13
INDOORS [EQ]	15	15
INNOVATION [ID]	5	5

POINTS ACHIEVED   POSSIBLE POINTS

Academic Achievement

A SCHOOL EXPANSION IN OUR NATION'S CAPITOL INTRODUCES A WETLAND TO A DENSE URBAN SITE

KEY PARAMETERS:

LOCATION: Washington, D.C., between Rock Creek and Glover Archbold watersheds

GROSS SQUARE FOOTAGE: 72,500 ft² (6,736 m²)

COST: $28.5 million

COMPLETED: September 2006

ANNUAL PURCHASED ENERGY USE (BASED ON SIMULATION): 19.4 kBtu/ft² (221 MJ/m²)

ANNUAL CARBON FOOTPRINT (PREDICTED): 4 lbs. CO_2/ft² (21 kg CO_2/m²)

PROGRAM: Classrooms, library, art/music rooms, science labs, constructed wetland, rooftop container garden

TEAM

ARCHITECT: KieranTimberlake Associates

COMMISSIONING AGENT: Engineering Economics

INTERIOR DESIGNER: Interior Design Resources

ENGINEERS: CVM Engineers (structural); Bruce E. Brooks & Associates (MEP); VIKA (civil)

LANDSCAPE: Andropogon Associates

ENVIRONMENTAL BUILDING: GreenShape

WETLAND CONSULTANT: Natural Systems International

GENERAL CONTRACTOR: HITT Contracting

In lieu of a traditional lawn, the courtyard at the heart of the middle school contains a constructed wetland that treats wastewater and manages stormwater.

HALKIN ARCHITECTURAL PHOTOGRAPHY

Most of the facility's conventional classrooms are retained within the original building, while the new wing offers science labs, art studios, and other special-purpose rooms.

Stephen Kieran, FAIA, notes that one of the biggest challenges his team faced was the aesthetic expectations that both the designers and the client brought to the project. "Some of the trustees had it in their heads that they could have a conventional brick Washington Georgian building and add features to achieve this level of performance," says Kieran. Instead, the project's green agenda led them to design fenestrations based on performance rather than a traditional aesthetic, to use wood from old wine vats as siding, and to devote the building's central courtyard to a constructed wetland rather than a lawn. Nonetheless, says Kieran, "With everybody working together, we reached agreement in the end."

The goal of managing wastewater on-site was accepted early, but the team's vision of how to do this evolved. "All through preliminary design, we were anticipating putting in a Living Machine," says Kieran, referring to a proprietary system in which wastewater is treated in a series of tanks, typically housed in a greenhouse. But regenerative-design consultant Bill Reed, AIA, argued that "a Living Machine is just another piece of equipment to fix a problem that we created." Reed suggested the constructed wetland that became the centerpiece of the courtyard.

Wastewater from the kitchen and bathrooms flows into settling tanks, where solids are collected before the water is released below the surface of the constructed wetland. After about 10 years, the solids will have to be removed to a landfill or composted, according to Reed. Surprisingly, city officials approved this alternative wastewater treatment system quickly. The city's health department had second thoughts at the last minute but ultimately agreed to let the project go

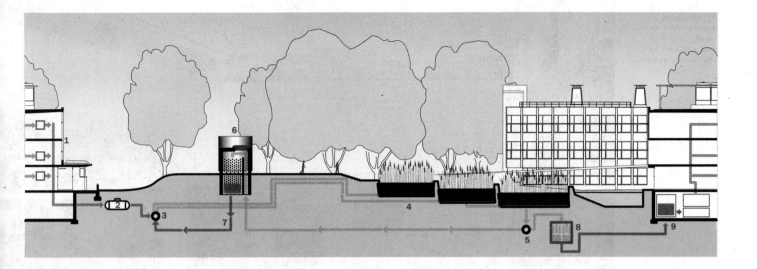

WASTE WATER TREATMENT SYSTEM

1 Wastewater from restrooms/laboratories 2 Pre-treatment tank 3 Flow splitter 4 Wetlands
5 Pump 6 Trickling filter 7 Return line 8 Sand filter 9 Reuse holding tank

ahead on a pilot basis. "We have a monitoring protocol that we have to follow," reports Saxenian. At press time, the wetland hadn't yet become fully operational.

The central wetland became the most prominent element in an integrated water-management system that begins with green roof areas that retain rainwater and also serve as garden space in which students grow vegetables for the cafeteria. With this approach, "the place is the process," notes landscape architect José Alminana of Andropogon of Philadelphia, and the enormous pedagogical value of the sustainability agenda became a driving force in the design process. In addition to the wetland, the designers introduced more than 80 plant species, all native to the Chesapeake Bay region. The biodiversity suggests that "the landscape becomes a new faculty member," says Alminana.

The control of water guided other design decisions. For example, the exterior cladding is a rainscreen system that includes a ventilated cavity to resist water intrusion. Interior finishes include cork, linoleum, bamboo, and wood flooring remilled from pilings extracted from Baltimore Harbor. In the landscape, flagstone was reused from sidewalks, and stone for walls came from a dismantled railroad bridge. Crediting the contractor's initiative in locating and scavenging the stone, Alminana notes, "A project of this ilk tends to attract this kind of thing. It doesn't happen by chance—the interest is contagious."

Energy-use reductions were achieved with a highly efficient building envelope, lighting controls, and passive strategies to minimize heating and cooling loads. Solar chimneys exhaust hot air during the cooling season without fans, and wind chimes in the towers signal airflow. Recognizing an opportunity to retire inefficient equipment in other buildings, the team designed the middle school's mechanical system to distribute hot and cold water to much of the campus.

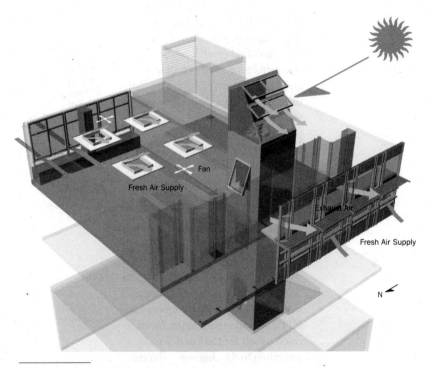

DIAGRAM

Solar chimneys designed for passive ventilation serve the specialty classrooms in the addition. South-facing glazing at the tops of the shafts heat the air within, creating a convection current that draws cooler air in through north-facing open windows. Portals in the shaftways within the building demonstrate the operation and effectiveness of the passive cooling systems with a telltale that moves with the breeze and a wind chime. The solar chimneys are also intended to be used in mechanical ventilation and air-conditioning modes, demonstrating the responsiveness of both passive and active systems to the local climate.

While Sidwell was built under a bid contract, Kieran argues that innovative projects are better procured through a negotiated contract. "You can't find enough bidders and contractors for LEED Platinum buildings that are willing to take all the risks," he says. With the contractors contributing to the design process, Kieran notes, they understand and buy into the importance of the green components. In this case, even before it won the bid, HITT Contracting was involved at certain points during the design phase to estimate costs and provide input on constructability, which improved the continuity between design and construction. Because the project had to be substantially completed during the school's 10-week summer break, HITT recommended using prefabricated panels for the exterior walls, according to the company's director of sustainable construction, Kimberly Pexton, AIA.

While HITT had previously constructed several LEED-certified projects, it entered new territory with several aspects of Sidwell, including photovoltaics

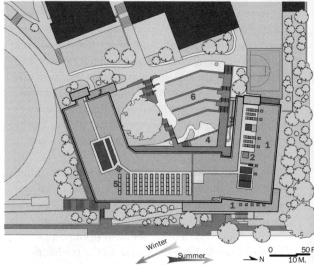

SITE PLAN

1 Green roofs
2 Solar chimneys
3 Water cascades
4 Biology pond
5 Photovoltaic panels
6 Constructed
7 Wetland

SKY CONDITIONS

Under the frequently overcast skies, a lot of glazing is needed to provide ample daylight.

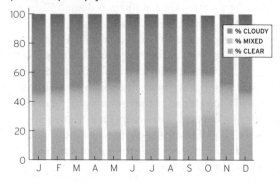

■ % CLOUDY
■ % MIXED
■ % CLEAR

and the constructed wetland. In both of these areas, dividing responsibilities among subcontractors was a challenge. If the photovoltaic (PV) provider isn't a licensed electrician, Pexton asks, "Where does the PV guy leave off and the electrician pick up?" For the wetland, the team thought it had found a subcontractor that could manage both the landscaping and the piping, but "when they got into it, the actual plumbing aspect was more than they could handle," says Pexton. So HITT turned that part of the work over to its plumber.

After extensive deliberation, the school elected to pursue LEED Platinum certification to serve as a beacon for the community, according to Saxenian. "We had some concern that this would be seen as frivolous, but we felt compelled by our core values and our belief in the importance of stewardship of natural resources," he says. Although the school will not prescribe a minimum LEED rating for future buildings, Saxenian says they expect their next project, a new lower school on the Bethesda campus, to achieve LEED Gold.

The school's commitment to using this project as a learning opportunity extends far beyond the students. A team from Yale University's School of Forestry and Environmental Studies is studying the school to determine if the project's green strategies have a measurable effect on student and faculty performance and health. But it will be harder to measure the long-term benefits of providing students with such a deep connection to natural systems, which is so rare in an urban setting. «

SOURCES

PREFABRICATED EXTERIOR WALL PANELS: Global Partners/ Symmetry Products Group

WINDOWS: Loewen Windows

DOORS: Algoma Hardwoods

LOW-SLOPE ROOFING: Sarnafil

MILLWORK: Greenbrier Architectural Woodwork

WALLCOVERINGS: Forbo Linoleum

ELEVATORS: Kone

INTERIOR AMBIENT LIGHTING: Finelite

CONTROLS: Lutron Electronics

BUILDING MANAGEMENT SYSTEM: Johnson Controls

FIRST FLOOR

1 Lobby
2 Offices
3 Classrooms
4 Library
5 Drama
6 Choral music
7 Instrumental music

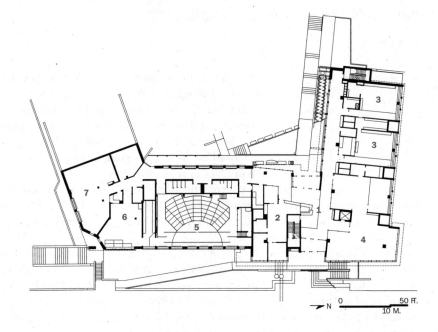

TEMPERATURES & DEW POINTS

The relatively high dew point (green line) reflects the region's high relative humidity.

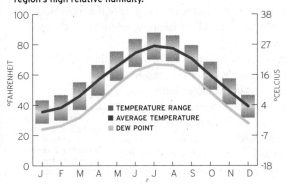

TEMPERATURE RANGE
AVERAGE TEMPERATURE
DEW POINT

HEATING/COOLING DEGREE DAYS

During the school year, heating is the primary load, but cooling is still needed during swing seasons.

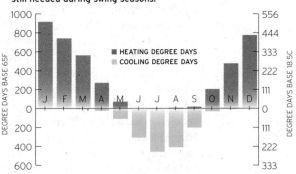

HEATING DEGREE DAYS
COOLING DEGREE DAYS

LEED SCORES
LEED-NC Version 2 Silver

SITES [SS]	8	14
WATER [WE]	3	5
ENERGY [EA]	10	17
MATERIALS [MR]	5	13
INDOORS [EQ]	7	15
INNOVATION [ID]	5	5

POINTS ACHIEVED   POSSIBLE POINTS

CASE STUDY

MERRILL HALL UNIVERSITY OF WASHINGTON, SEATTLE

Rebirth and Regeneration

A HORTICULTURE SCHOOL'S COURTYARD ATRIUM BREATHES NEW LIFE INTO A RUINED SITE

KIRA GOULD

HE FIRST MERRILL HALL, HOME TO THE CENTER FOR URBAN HORTICULTURE at the University of Washington (UW), was destroyed in 2001 by a fire set by Environmental Liberation Front eco-activists. Their research target wasn't housed at Merrill, but its world-class archives, library, and herbarium were decimated. The loss of that original building, designed by Jones & Jones, and much loved by the center's staff, faculty, and community residents, made the urgency of rebuilding greater.

The center is situated at the edge of campus, between a residential neighborhood and a nature preserve. The Merrill complex includes the Elisabeth C. Miller Library, the Hyde Herbarium, academic offices, and labs. The academic and outreach programs run by the center attract some 65,000 people a year.

Professor Tom Hinckley was director of the center at the time of the fire and throughout the rebuilding project. "As soon as we started talking about rebuilding, we were talking about sustainability," he recalls. The staff and faculty were inspired by a video about the Chesapeake Bay Foundation headquarters, but the university initially resisted, reportedly worried that going green would set a costly precedent. The center convinced the university, but was told that it would have to raise all the money for sustainability equipment, materials, and systems.

Arborist Sue Nicol was the center's outreach coordinator and served as its representative on the project team. "We wanted to make the tragedy of the fire into something meaningful," Nicol says. "We felt that the best way to do that would be to make the buildings match the values of the center."

For Craig Curtis, of Miller|Hull Architects, the idea of the facility being the edge between two conditions was inspiring. "We wanted to create a building that itself was part of the transition from urban to rural," he says. Because of the tight budget and the desire to rebuild quickly, the program stayed the same, with a similar footprint as well.

The greenhouse, "Merrill Commons," which opens to the lobby of the new building off of its main entrance, is a new and much-needed informal gathering space. Celebrating the entry was an important goal, since one weakness of the old facility was a somewhat undefined set of entry points. Early on it was determined that passive ventilation would be sufficient for cooling most spaces although not the library. High-efficiency condensing boilers and a water-cooled chiller were selected to keep energy use low, and administrative staff agreed to open offices for the light and air benefits. "It works well," Nicol says, "but it couldn't have happened with-

>> Merrill Hall's new greenhouse creates an informal gathering place at the entrance, providing a gradual transition from indoors to out.

SOURCES

ROOFING:
American Hydrotech Green Roof; Ballard Sheet Metal Energy-Star metal roofing

FLASHING, ACCESSORIES:
AEP-Span Klip-Rib

SALVAGED LUMBER PANELING:
Matheus Lumber

TABLES FROM SALVAGED TREES:
Urban Hardwoods

COUNTERTOPS:
Richlite

LIGHTING FIXTURES:
KIM Lighting

CHILLER: Airstack, with 2-ASP 15 modules

PLUMBING: Falcon Waterfree urinals, Caroma Walvit dual-flush toilets, Chicago Faucets 650-4 low-flow faucets

out staff buy-in and training." Sarah Reichard, who manages two labs and the herbarium, and helped raised money for the project, says she has been surprised at the natural ventilation's effectiveness. "We have high windows in the offices that stay open full-time in warm weather, as well as transom windows. It makes a noticeable difference to have these open."

The center is part of the College of Forestry Resources, which is supported in part by timber companies championing the Sustainable Forest Initiative (SFI), so its use of wood was important. Instead of opting solely for Forest Stewardship Certified (FSC) wood, as LEED suggests, the design team decided to embrace and display the debate within the forestry community over FSC versus SFI.

The city of Seattle provided a 9.6-kW photovoltaic array as a demonstration component, which provides about 9 percent of the facility's needed energy. The library's meter shows energy generation in real time; occupants note the system is performing 10 percent better than expected.

The building forms and the large courtyard are designed to restore hydrological flows. Indoors, low-flow fixtures were utilized where possible; predicted annual use is 51,675 gallons (compared to a baseline case of 81,738). The site includes an existing rainwater cistern to which the roof and site water are channeled. The whole-site water system links to an existing bioswale that had been in place for two decades; Merrill is the first campus building to link to it. As it turned out, the low-flow fixtures were one of the big lessons of the project. The angle of the pipes leading to the sewer system from dual-flush toilets was not great enough to promote the flow with the lesser amount of water, and they wound up being replaced.

Another lesson involved the university's decision not to specify low-flow hoods in the labs. The less-efficient models are noisy and some are being replaced. And mechanical units on the roof are louder than expected, which interferes with some events held in the courtyards. "Acoustic comfort is a big issue and I think this is something that we will be looking at more closely," Curtis says of his firm's future projects. "We are asking these questions much earlier and are thinking about consulting with additional people."

Project manager Norm Menter of UW acknowledges that the building doesn't work perfectly, but he's not perturbed. "Adjusting parameters means that we have to retest assumptions," he says. He credits the passion of the client for the success of the project. "The team at the center believed that a smaller footprint was possible, and necessary. They felt the building could be an asset to their program and to the educational process that they nurture—learning how urban and natural environments relate. This building is that process in real time." «

SKY CONDITIONS
The generally overcast conditions make protected, outdoor gathering spaces all the more popular.

TEMPERATURES & DEW POINTS
Seattle's relatively mild temperatures make natural ventilation a viable cooling strategy for most spaces.

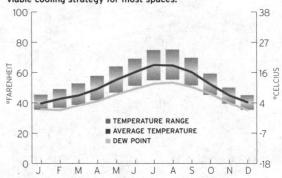

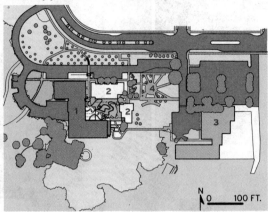

SITE PLAN

1 New project 2 Existing conference
3 Existing greenhouse 4 Gardens

N 0 — 100 FT.

⤊
Users were very supportive of the idea of relying on natural ventilation. Special funding was secured for airflow modeling to show that temperature requirements could be met.

SECTION WITH WINDFLOW DIAGRAM

1 Cross ventilation
2 Stack ventilation

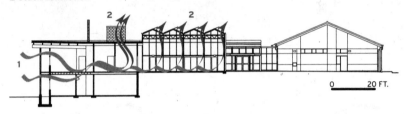

0 — 20 FT.

«
With extensive public use of the herbarium, the facility serves as a learning tool for students from many university departments.

ENTRY LEVEL

1 Entry 2 Library 3 Herbarium 4 New commons greenhouse
5 Existing courtyard 6 Demonstration green roof 7 New rain garden

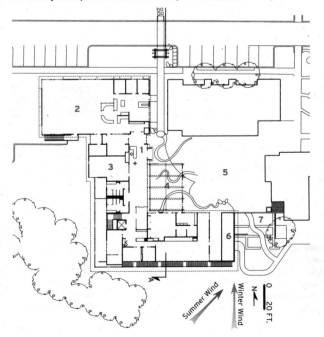

Summer Wind   Winter Wind   N   0 — 20 FT.

HEATING/COOLING DEGREE DAYS

Seattle's climate requires more heating than cooling for small buildings that don't have a lot of internally generated heat.

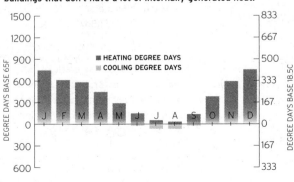

- HEATING DEGREE DAYS
- COOLING DEGREE DAYS

S is for Sust

THOMAS L. WELLS PUBLIC SCHOOL, TORONTO

LEED SCORES
LEED-Canada NC 1.0 Silver

SITES [SS]	8	14
WATER [WE]	2	5
ENERGY [EA]	3	17
MATERIALS [MR]	5	14
INDOORS [EQ]	14	15
INNOVATION [ID]	5	5

■ POINTS ACHIEVED ■ POSSIBLE POINTS

KEY PARAMETERS

LOCATION: Toronto, Ontario, Canada (Rouge River watershed)

GROSS SQUARE FOOTAGE: 71,194 ft²/ 5,554 m²

COST: $16 million

COMPLETED: August 2005

ANNUAL ENERGY USE (BASED ON SIMULATION): 76 kBtu/ft² (860 MJ/m²)– 35% reduction from base case

ANNUAL CARBON FOOTPRINT (PREDICTED): 22 lbs. CO_2/ft² (108 kg CO_2/m²)

PROGRAM: Classrooms, offices, cafeteria, gymnasium, library

THOMAS L. WELLS TEAM

OWNER: Toronto District School Board

ARCHITECT: Baird Sampson Neuert architects

COMMISSIONING AGENT: The Mitchell Partnership

ENGINEER: Blackwell Bowick Partnership Ltd. (structural); Keen Engineering, now part of Stantec Consulting (mechanical); Mulvey & Banani International (electrical)

LANDSCAPE ARCHITECT: Elias + Associates

LIGHTING: Mulvey & Banani International

ACOUSTICAL: Aercoustics Engineering

BUILDING SCIENCE/SUSTAINABILITY: Ted Kesik

GENERAL CONTRACTOR: Struct-Con Construction

ainability

A GREEN PHILOSOPHY SETS PRECEDENT FOR SUBURBAN SCHOOL DISTRICT'S EXPANSION PLANS

JESSICA BOEHLAND

HE TORONTO DISTRICT SCHOOL board had the right idea from the start. "They said, 'don't give us a bunch of the green design icing; give us the cake,'" says Seth Atkins, associate at Baird Sampson Neuert architects and project coordinator for the Thomas L. Wells Public School outside Toronto. "They said, we don't want things that read as green design but don't have a big effect. We want less green roof and more in terms of high-efficiency boilers, heat recovery, and high-performance glazing." The resulting building, which opened in time for the 2005–2006 school year, does have a rooftop garden, but its green design sparkles mostly through less flashy features.

The two-story, 71,000-square-foot Wells School sits on three acres in the midst of a new housing development in Scarborough, a fast-growing Toronto suburb home to many Asian and Middle-Eastern immigrants. It was designed to serve 670 students in kindergarten through grade eight.

Wells was the Toronto school district's first venture into green design, according to David Percival, an architect who serves as the district's manager of standards compliance and environment. The school board, however, which manages nearly 600 facilities, intended Wells' green philosophy to set a precedent for future schools. The request for proposals stressed a desire for an integrated design process and an energy-efficient building with good indoor air quality. Once Baird Sampson

⌃
A bench that runs along the school gymnasium offers students a resting place while they wait to be picked up at the end of the day. Homes crowded into this residential development are reflected in the windows.

Neuert architects had been selected to lead the project, the board hosted several design charrettes, which involved everyone from designers and consultants to teachers, maintenance staff, and community members.

Midway through contract documents, the architects indicated the working budget was insufficient to meet the project goals, so the board increased the budget by nearly 10 percent, to $12.6 million Canadian. This allowed the team to consider green alternatives that would pay for themselves after about 10 years, opening the door to innovations, such as heat-recovery ventilation, building automation, and radiant heating and cooling, that might not have been feasible within a more conventional budget.

Since the school board plans to operate Wells for at least 75 years, the team selected durable materials that would require little maintenance. "We considered durability and longevity of systems to be critical to sustainability," says Percival. Material choices included suspended gypsum wallboard ceilings in place of less-durable acoustic tile, for example, and porcelain tile flooring,

which is easy to clean with mild soap and doesn't need to be stripped and waxed, in place of standard vinyl composition tile. Additionally, the project team used low-VOC materials throughout the project.

The team was also concerned about energy efficiency. "How to use a lot of glass for daylighting yet still get energy reduction challenged our design team," says Atkins. "For us, that meant utilizing thermal mass," which absorbs heat during the day and releases it at night, reducing internal temperature swings and saving energy. This focus on passive solar design drove the building's orientation and layout. Classrooms face south to harvest daylight. Much of the glazing is recessed in the masonry building envelope and furnished with exterior lightshelves to shade the windows in the summer when the sun passes high overhead, and bounce daylight deep into the rooms in the winter, when the sun sits lower in the sky. All of the school's classrooms provide daylight and views. Even the gymnasium is washed in even, diffuse daylight.

The school board had the right idea: "Don't give us the green design icing; give us the cake."

The project's ventilation system is probably its most innovative feature. "We had three people proposing three different systems," Atkins says, until the design team realized the disparate systems could provide superior performance when effectively combined. In the resulting fusion, ventilation air is supplied to rooms at floor level near the corridors. The air moves slowly across the floor and up along the windows to grilles in the ceiling. Then, instead of passing through ductwork above a dropped ceiling, the air moves directly through the hollow-core precast concrete slabs to ducts in the corridors. As it moves through the slabs, the return air picks up heat. In the summer that heat is expelled, and in the winter it is captured for reuse. Pleased with the results, Baird Sampson Neuert hopes to patent the system.

While the design team used both the LEED rating system and the Collaborative for High Performance Schools (CHPS) guidelines as design tools, the school board elected not to attempt LEED certification. During construction, however, it changed its mind. The team originally registered the project through the U.S. Green Building Council but shifted to the Canada Green Building Council once it became established. The school earned a LEED Canada Silver rating in September 2006.

⌃
The school's south facade features extensive glazing and exterior light-shelves that allow daylight to reach deep into the classrooms, while interior clerestories spread light from corridors into adjoining rooms.

«
The Thomas L. Wells Public School sits amidst a dense housing development in the Toronto suburb of Scarborough.

SITE PLAN

1 School
2 Soccer field
3 Creek
4 River

SKY CONDITIONS

The largely overcast winter conditions drove the project team to maximize daylighting for student well-being.

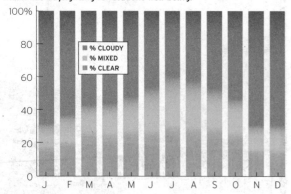

- % CLOUDY
- % MIXED
- % CLEAR

HEATING/COOLING DEGREE DAYS

Toronto's cold climate illustrates the importance of an efficient heating system.

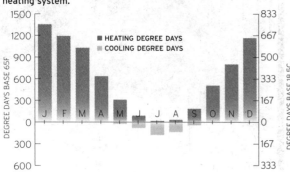

- HEATING DEGREE DAYS
- COOLING DEGREE DAYS

TEMPERATURES

The relatively large difference between high and low temperatures in the summer facilitates passive cooling.

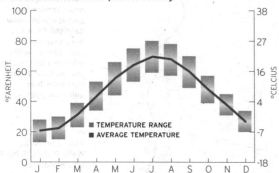

- TEMPERATURE RANGE
- AVERAGE TEMPERATURE

Since the school board plans to operate Wells for at least 75 years, the team selected durable materials that would require little maintenance.

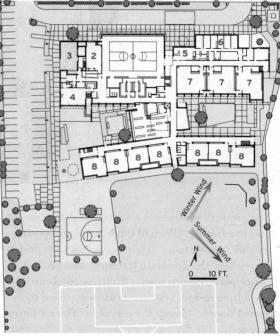

ENTRY LEVEL

1 Gym
2 Stage
3 Service
4 Staff
5 Administration
6 Offices
7 Kindergarten
8 Classrooms

Winter Wind

Summer Wind

N

0   10 FT.

Upon completing the project, the team carried out a comprehensive post-occupancy review, including a survey of all faculty and staff members. On a score of one to five, the overall ratings came back well above four. "The ones that really hit high marks were daylighting and views," says Atkins. "They had fives straight across." The only area to come in below four was acoustics, which the team has since addressed by installing acoustically absorptive wallboard in key transmission areas.

The project faced a few other difficulties as well. Percival noted that while the displacement ventilation works well in most of the school, it can be too noisy in the gym. "It's only a problem when they're having assemblies or other large gatherings," he says. Maintenance personnel have struggled with the ceramic tile. While it is easier to care for than vinyl, the grout

HEATING & COOLING DIAGRAM

1 Durable tile flooring reflects light 2 Radiant floor heating provides even heat distribution 3 Top and bottom operable vents allow passive cooling 4 Thermal mass stores solar energy in winter and slows heat transfer in summer 5 Return air moving through hollow-core slab picks up or dissipates heat 6 Corridor lighting penetrates into classrooms through transom 7 Return air duct completes the loop to the air handler

SOURCES

STRUCTURAL STEEL: Vicwest

CAST-IN-PLACE CONCRETE: Coreslab

CLADDING: Hanson Brick

CURTAINWALL LOW-E INSULATED GLAZING UNITS: Triple-seal doors by Doorlam

WASHROOM ACCESSORIES, CABINTWORK, AND CUSTOM WOODWORK LOW-VOC CONTENT: Bobrick Washroom Equipment

CARPET: Interface

between the tiles is difficult to clean, and chairs and desks have scuffed the tile surfaces. Atkins believes terrazzo flooring would have been preferable. Ruth Jory, principal at Wells school, reported that the classrooms are sometimes too bright, but that window film and blinds have remedied the problem.

The post-occupancy review also indicated that the school could be operated more efficiently. As a result, Atkins spent two weeks observing how staff members were using the building and working with them to develop more efficient habits. He explained, for example, how turning off the mechanical ventilation when they opened the windows saves energy. "It completely changed their behavior," he says. Jory agrees. "We have embraced the green philosophy that was inherent in the design," she says, "and look forward to ensuring that future students and staff continue to preserve our energy-efficient school."

Though the students are unaware of most of the school's green features, they love the space. Atkins notes that the green design, and especially the daylighting, "seems to spark a curiosity in them." Percival says he hopes that the teachers will incorporate some of the building's green aspects into their lesson plans.

Before the school even opened its doors, the student body was spilling over Wells's capacity. "Once it got out that this school would have improved indoor-air quality and daylighting, everyone who could pulled their kids from other schools and enrolled them in Wells," says Atkins, resulting in the temporary use of portable classrooms. In response, the school board is planning another K–8 school about a mile north of Wells. It too will be green. «

Chapter 4

GOVERNMENT

Government bodies have been some of the strongest advocates for sustainable building in the United States by mandating green standards in new buildings and renovations. For example, the General Services Administration (GSA), an agency which procures and manages facilities for most federal agencies and, with over 8,000 owned and leased buildings, is the nation's largest landlord, required in 2003 that its new building projects meet LEED-certified requirements, with a target of LEED Silver. Other government bodies throughout the United States, from the federal to state and municipal levels, have similar mandates. The three case studies in this chapter were selected as exemplars of sustainable, government-owned buildings.

Even without the GSA mandate, the National Oceanic and Atmospheric Administration, seeing itself as a steward of the environment, would have aimed for its new Satellite Operations Facility in Suitland, Maryland, to be sustainably designed. The GSA pushed the project further by enrolling it in its Design Excellence Program, which, through a two-step architect-engineer selection process, promotes innovative design. A model of good design sense fused with sustainable goals, the building features a vegetated roof that merges seamlessly with the landscape to the north, lightening the visual weight of the large building.

The California Public Employees' Retirement System's headquarters complex, in Sacramento, went beyond the state mandate of LEED Silver to achieve LEED Gold through an ambitious approach to high-tech sustainable design. The recovery of waste heat from chillers to make hot water, along with many other features of the building's mechanical system, made this possible. Relative to a comparable building designed to merely comply with ASHRAE 90.1-1999, the building uses 38 percent less energy.

Missouri does not require its government buildings to be LEED certified, but the state's history motivated the architects of the Lewis & Clark State Office Building, in Jefferson City, to push the limits of green design. Inscriptions on the state capitol building describing the state's abundant natural resources made clear the importance that civic forefathers had placed on the environment. Not far from the building's site is the location from which Lewis and Clark embarked on their trans-American trek, a fact that further inspired the architects. The design team achieved a LEED Platinum rating in part through a very efficient daylighting scheme and a 50,000-gallon cistern that collects rainwater to flush toilets throughout the building. Although a great boon for sustainable building, state mandates can only go so far. These projects demonstrate that going beyond the requirements can often be the best public service a government body can provide. «

The center of the building is a four-story atrium, offering access to the outside. Occupants generally prefer the stairs to the elevators.

KEY PARAMETERS

LOCATION: Jefferson City, Missouri (Missouri River watershed)

GROSS SQUARE FOOTAGE: 120,000ft^2 / 19,300 m^2

COST: $18.1 million

COMPLETED: March 2005

COST: $18.1 million

ANNUAL ENERGY USE (based on utility bills): 68 kBtu/ft^2 (775 MJ/m^2)

ANNUAL CARBON FOOTPRINT: 20 lbs. CO_2/ft^2 (97 kg CO_2/m^2)

PROGRAM: Offices, conference rooms, food service, atrium

LEWIS & CLARK TEAM

OWNER: State of Missouri, Missouri Department of Natural Resources

ARCHITECT AND INTERIOR DESIGN: BNIM Architects

COMMISSIONING AGENT: Sys-Tek

ENGINEER: Structural Engineering (structural); Smith & Boucher (mechanical); FSC (electrical); SK Design Group (civil)

LANDSCAPE: Conservation Design Forum

ENVIRONMENTAL BUILDING DAYLIGHT DESIGN/ENERGY STRATEGIES: ENSAR Group (now RMI/ENSAR Built Environment)

LIGHTING: Clanton Engineering

GENERAL CONTRACTOR: Professional Contractors and Engineers

MECHANICAL SYSTEMS DESIGN: Rumsey Engineers

COST ESTIMATOR: Construction Cost Systems

CASE STUDY
LEWIS & CLARK STATE OFFICE BUILDING JEFFERSON CITY, MO

New Frontiers in Office Space

STATE OFFICE BUILDING ENRICHES EMPLOYEES' LIVES WITH LIGHT AND LANDSCAPE

TRISTAN KORTHALS ALTES

VEN BEFORE THE OUTSET OF DESIGN AND CONSTRUCTION of the Lewis and Clark State Office Building, the mission was clear: create a green office building and certify it to the highest level of LEED without overtaxing the credulity of the taxpayers in the "Show-Me States." The modest budget, $18.1 million, was provided by the Missouri state legislature before the tenant, the Department of Natural Resources, had even put the sustainability goals on the table. But it was the "writing on the wall" that crystallized the building's environmental goals.

"When we were interviewing for the project, we spent one afternoon in the state capitol building reading inscriptions that are carved into a particular wall about the natural resources of the state," says architect Steve McDowell, FAIA, of BNIM Architects. "We read about the minerals, animals, plants, rivers—all the natural things that were important to the character and nature of Missouri," he says. "We thought we were already committed to the green agenda, but looking back at what was important to the founders and early citizens of the state influenced how deeply we held those values through the project." That commitment was nurtured from the beginning of the process, when BNIM organized a design char-

MIKE SINCLAIR PHOTOGRAPHER

LEED SCORES
LEED-NC Version 2 Platinum

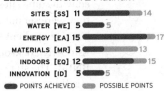

	POINTS ACHIEVED	POSSIBLE POINTS
SITES [SS]	11	14
WATER [WE]	5	5
ENERGY [EA]	15	17
MATERIALS [MR]	5	13
INDOORS [EQ]	12	15
INNOVATION [ID]	5	5

SITE PLAN

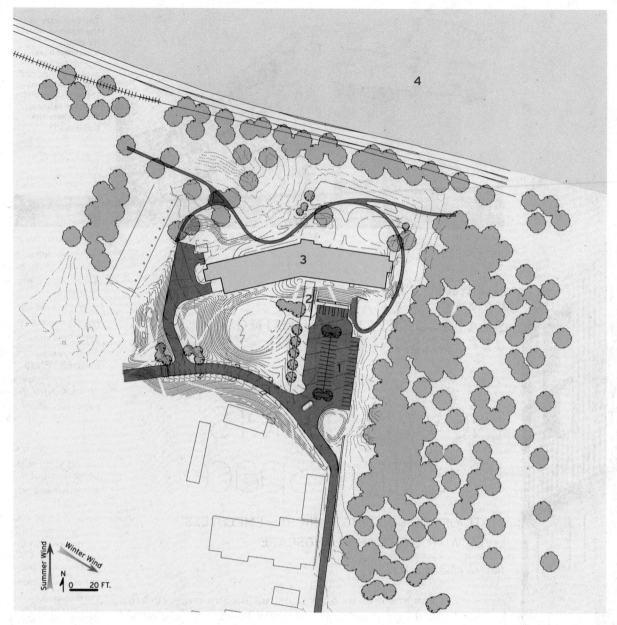

rette that involved more than 100 people, representing all parties involved in the project.

Not far from where Meriwether Lewis and William Clark embarked on their exploration of the American West in 1804, the building is a stone's throw from the Missouri River. The 120,000-square-foot structure extends about 350 feet along the east-west axis, and only 70 feet from north to south, a ratio that was calculated to reduce energy costs and maximize interior daylight.

Choosing the site was integral to the project. "We started with some sites that we felt were unsuitable, because they were set out in the suburbs, for example, so we challenged our client to look for a more urban site," says Kimberly Hickson, AIA, one of the project managers for BNIM. The state came back with 17 sites, including the one that was eventually chosen.

Constructed on the 144-acre site of the former Jefferson City Correctional Facility, eight blocks from downtown, the office building is one piece of an ambitious mixed-use urban redevelopment project. This building took the place of a former women's prison, with bricks from the prison being used for a number of interior surfaces.

In a landscape design based on xeriscape principles, indigenous grasses, shrubs, and trees have thrived despite a dry first year. Vegetated bioswales and topography that encourages stormwater infiltration help meet a goal of keeping runoff out of the municipal stormwater system. The landscaping is not without maintenance needs, but they are considerably less than in a conventional landscape. "We pull weeds from it. It has been mowed once," says Dan Walker, director of the general services program for the Department of Natural Resources, who represented the tenant throughout

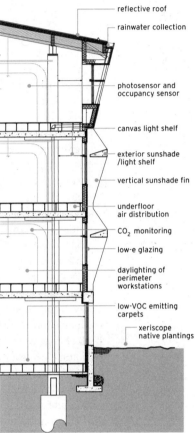

«

Horizontal lightshelves and vertical fins on the exterior facade, both of concrete, block direct sunlight while reflecting it into the buidling's interior. Open offices are located on the building's perimeter, where they enjoy this daylighting, while building services are located at the core.

SECTION

reflective roof

rainwater collection

photosensor and occupancy sensor

canvas light shelf

exterior sunshade /light shelf

vertical sunshade fin

underfloor air distribution

CO_2 monitoring

low-e glazing

daylighting of perimeter workstations

low-VOC emitting carpets

xeriscope native plantings

design and construction. Nature trails around the site and reaching toward the river are planned.

Green features extend from the outside into the building. A 50,000-gallon cistern collects rainwater from the roof, which is filtered and used in flushing toilets. The system conserved 405,000 gallons in its first 13 months. One hundred and sixty-eight photovoltaic panels produce 21.5 kW of power, or 2.5 percent of the building's needs. Thanks in large part to a well-planned daylighting system and the careful design of the heating and cooling system, the building is predicted to use less than half of the energy of a comparable ASHRAE base model.

The narrow aspect ratio of the building, solar orientation, and both interior and exterior lightshelves help daylight penetrate deep into the office space. Employees enjoy access to views and daylight at workstations that are located around the perimeter. Enclosed rooms are

generally situated at the core. Many of the windows are operable, providing natural ventilation. Although the daylighting scheme went through computer modeling by ENSAR Group, which has since merged with the Rocky Mountain Institute's Green Development Services, during part of the winter a gap between the interior canvas lightshelves and the glazing causes glare in some workstations. A miscommunication with the fabricator led to the mistake, which wasn't present in models, and the problem was fixed in the winter of 2007.

The building design also keeps employees active. Amenities such as changing rooms and showers were created largely for those who bike to work, but also encourage occupants to exercise. A number of people jog during their lunch breaks, according to Walker, and some walk downtown for meetings. Nearly everyone uses the prominent and attractive stairs in the central atrium.

SOURCES

BICYCLE RACKS:
BRP Enterprises
WA2-11-SM

UNIT MASONRY:
Prairie Stone
Northfield Block

**STRUCTURAL
GLUED-LAMINATED
TIMBER:**
Mississippi
Laminators

**EXTERIOR
ARCHITECTURAL
WOODWORK:** Trex
(entry sun
shade)

LINOLEUM:
Armstrong
Colorette

**BUILDING
INSULATION:**
Bonded Logic
cotton fiber;
International
Cellulose spray
applied

SIDING: James
Hardie Siding
Hardiplank lap
siding

INTERIOR DOORS:
VT Industries
Curries
quartersawn
ash

**ALUMINUM
ENTRANCE AND
WINDOWS:**
Kawneer

**TUBULAR
SKYLIGHTS:**
Huvco

GLAZING: Low E
Insulating Glass,
Viracon VE
Ceramic-coated
spandrel
Insulating glass,
Viracon Heat
Mirror

**LIGHTING
CONTROLS:**
Lutron

PV: Uni-Solar
PVL Field
Applied
Laminate

**LOUVERS AND
VENTS:**
Greenheck

RAISED FLOOR:
Haworth
TecCrete XL

**FABRIC
LIGHTSHELVES:**
Rosebrand
polyester/lycra

Although all of these features reduce operating costs and increase occupant satisfaction, the building team was still concerned that its construction budget would lead to compromises. At one point, the state balked at buying the low-emissivity glass specified by the architect, so the firm ran an energy model with a lower-cost alternative. "We found that we could spend the money on glass or we could spend it on a bigger mechanical system," Hickson says, noting that the mechanical system would add energy and maintenance costs over the long run. The glass stayed. The engineer performed a similar analysis on the building's aspect ratio when the client wanted a boxier profile; again, the analysis confirmed the efficiency of the design.

The project used an integrated design process, but owing to the requirements of competitive bidding, a contractor wasn't involved early on. "I guess we were all somewhat nervous at the bid opening," says Hickson. "We didn't really know where it would fall." The team received a low bid from a trusted contractor, although one without experience in green building. The contractor worked hard to meet the sustainability goals and helped the team gain LEED points in construction waste recycling and recycled content of materials, at a time when the project was on the edge of achieving Platinum. "There was a lesson there," says Laura Lesniewski, AIA, BNIM's project manager during the construction phase. "If you have someone who's interested in learning, he doesn't need to have prior experience."

Despite the project's impressive achievement in the LEED rating system, the team gave up points along the way. It procured lumber from Missouri's only certified sustainable forest for the atrium, but for the roof structure, which uses exposed glued laminated beams, the team couldn't locate a product with certified content. It also hoped to earn a point for reducing the urban heat-island effect with reflective roofing, but the emissivity level of the roof left them a fraction short of the requirement.

The team went to extra lengths to achieve some points, such as using furniture that meets indoor air quality (IAQ) standards. State agencies are required to purchase furniture manufactured through prison inmate vocational programs, so the team brought its sustainability agenda to the prisons, working to help them achieve Greenguard Indoor Air Quality listing for its furniture.

"The building performs well and our occupants are extremely happy from the perspective of IAQ, lighting, and heating and cooling," says Walker. Several occupants have reported better health in this building after having sinus problems in previous buildings, he adds. Walker has also noted reduced absenteeism, an observation the department plans to investigate by examining employment records.

By all accounts, the success of the project was based on the commitment to sustainability by the key members of the team. Built on budget, the building is also meeting financial performance expectations over the long term. "We did life-cycle costing on every aspect of the building, and so far things are on schedule," says Walker. «

⌃
**Certified
sustainable lumber
from Missouri's
only Forest-
Stewardship-
Council-certified
forest was used in
the atrium, which
is filled with
daylight and views
of the bluffs.**

»
**A non-irrigated
landscape design
with indigenous
grasses, shrubs,
and trees has
thrived.**

SKY CONDITIONS

This chart illustrates the average proportion of cloudy, mixed, and clear sky conditions over the course of a year.

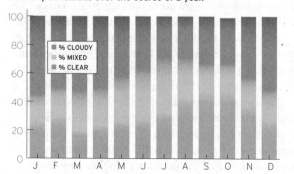

SECTION PROFILE

SECTION A-A

1 Entry
2 Atrium

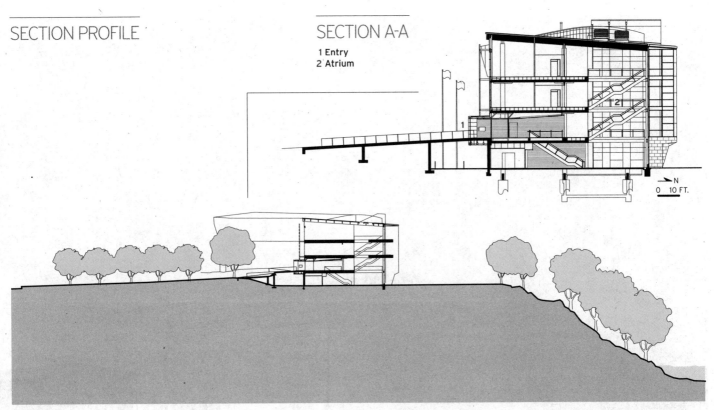

N
0 10 FT.

FLOOR PLAN

1 Entry
2 Atrium
3 Open office
4 Teaming area
5 Recycling
6 Interview
7 File/storage

N
0 20 FT.

TEMPERATURES & DEW POINTS

The blue bars show normal monthly high and low temperatures over thirty years; extremes on any given day can be much higher or lower.

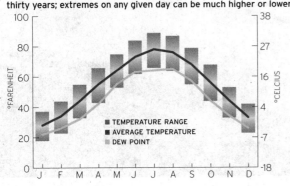

°FARENHEIT

°CELCIUS

■ TEMPERATURE RANGE
■ AVERAGE TEMPERATURE
■ DEW POINT

J F M A M J J A S O N D

HEATING/COOLING DEGREE DAYS

Jefferson City, Missouri, has a significant need for heating, as well as moderate need for cooling.

DEGREE DAYS BASE 65F

DEGREE DAYS BASE 18.5C

■ HEATING DEGREE DAYS
■ COOLING DEGREE DAYS

J F M A M J J A S O N D

Close to Home

A HEADQUARTERS CONSOLIDATION EXCEEDS THE SUM OF ITS PARTS

CALPERS HEADQUARTERS COMPLEX SACRAMENTO, CALIFORNIA

JESSICA BOEHLAND

THE CALIFORNIA PUBLIC EMPLOYEES' Retirement System (CalPERS) is the nation's largest public pension fund, with 1.5 million members, more than 2,500 employees, and an investment portfolio valued at more than $220 billion. So, when it decided to expand its Sacramento, California, headquarters, the organization thought big. The resulting project, covering two downtown city blocks, includes 550,000 square feet of office space, 25,000 square feet of retail space, and parking for 1,000 cars. Three residential developments totaling another 180,000 square feet are either underway or planned for nearby sites.

CalPERS took its design inspiration from its existing home, Lincoln Plaza North, which was completed in 1986. Featuring raised floors, extensive daylighting, and 180,000 square feet of roof terraces, Lincoln Plaza had convinced the organization of the potential value of green design. CalPERS wanted the expansion to complement Lincoln Plaza while projecting an image of stability and permanence, providing a productive and comfortable work environment, and creating a lasting and meaningful contribution to the organization and community.

"From the very beginning, CalPERS saw this as their home," says Anthony Markese, AIA, design principal at Pickard Chilton Architects of New Haven, and the organization treated the expansion as a long-term investment. Both the CalPERS chief of plant operations and the project's green building consultant were embedded in the design process from the beginning, says Diana Proctor, project manager at CalPERS. That involvement helped the team maintain "a consciousness of operations and maintenance" throughout the design and construction process, according to Proctor.

LEED SCORES
LEED-NC Version 2 Gold

	Points Achieved	Possible Points
SITES [SS]	6	14
WATER [WE]	1	5
ENERGY [EA]	11	17
MATERIALS [MR]	6	13
INDOORS [EQ]	11	15
INNOVATION [ID]	5	5

● POINTS ACHIEVED   ● POSSIBLE POINTS

KEY PARAMETERS

LOCATION: Sacramento, California (Sacramento River watershed)

GROSS SQUARE FOOTAGE: 1.1 million ft² (102,000 m²)

COST: $192 million

COMPLETED: November 2005

ANNUAL PURCHASED ENERGY USE (BASED ON SIMULATION): 81.6 kBtu/ft² (927 MJ/m²), 16% reduction from base case (65% of the energy use is for data center and office equipment)

ANNUAL CARBON FOOTPRINT (OFFICE AND DATA CENTER ONLY) PREDICTED: 20 lbs. CO_2/ft² (100 kg CO_2/m²)

PROGRAM: Office, retail, housing (proposed), underground parking

TEAM

OWNER: California Public Employees' Retirement System (CalPERS)

ARCHITECT: Pickard Chilton

ARCHITECT OF RECORD: Kendall-Heaton & Associates

ASSOCIATE ARCHITECT: Dreyfuss & Blackford Architectural Group

INTERIOR DESIGNER: IA Interior Architects

LANDSCAPE: Hart-Howerton

ENGINEERS: Nolte & Associates, Inc. (civil), Carter & Burgess and CYS Structural Engineers (structural), Arup (MEP, facade)

COMMISSIONING AGENT: Capital Engineering

ENVIRONMENTAL CONSULTANT: Arup with Simon Associates

CONSTRUCTION MANAGEMENT: Turner Construction

The organization planned the new project for a rectangular site with the long sides facing north and south, a boon for daylighting. At more than a million gross square feet, however, the project's sheer size threatened its ability to capitalize on this orientation. In response, the team broke the project into two U-shaped buildings—one four stories tall and the other six—facing one another. This decision not only permitted daylight to reach a greater percentage of the interior space, but it also accommodated an existing street bisecting the site and allowed for a public courtyard inside the donut-shaped building.

The team then designed the interior to make the most of this daylighting potential. Thanks to its raised floors, the project features floor-to-ceiling heights of 10-feet 8-inches, about 2 feet taller than usual, allowing more light to enter each floor. Additionally, the team located enclosed rooms at the interior of the relatively thick bases of the U-shaped buildings, devoting the perimeter to open spaces. "It was a more democratic way to share light and views," says Markese. Paired with facade-integrated planters supporting native grasses, exterior and interior lightshelves provide shade, reduce glare, and bounce daylight deep into the building.

CalPERS was designed to use 38 percent less energy than a comparable building designed in minimal compliance with ASHRAE 90.1-1999. Its energy-efficient features include not only daylighting, but also glazing with a U-value of 0.26 to 0.29 and a solar heat-gain coefficient of 0.38, operable windows in some areas, underfloor air distribution, and the recovery of waste heat from chillers to make domestic hot water. A platform on the building's roof shades mechanical equipment

» The CalPERS facade blends vertical and horizontal elements, including exterior and interior lightshelves, to filter and shade the sunlight, while allowing for daylight penetration and views from the building's interior.

Replacing surface parking with below-grade parking

while also supporting an 87 kW photovoltaic (PV) array. "We were quite lucky in that SMUD [the Sacramento Municipal Utility District] had a system in place where, if you provided the structure, they provided the PV panels," says Markese, "and CalPERS had the foresight to say 'let's take advantage of this.'"

Among the design team's most significant decisions was eliminating on-site surface parking. While some parking is available under an elevated highway less than a mile from the site, the majority is located on two levels under the complex. The strategy "really transformed the building," says Markese. "It provided the freedom to sculpt the building, to create spaces that would be a gift to the city, and to not have to deal with the empty facades of a parking garage." It also reduced the project's development footprint and its contribution to the urban heat-island effect.

When the design team began work in 1999, the U.S. Green Building Council's LEED Rating System was in its infancy. "This was definitely the first LEED project that anyone on the team had worked on," says Markese, noting that "the precepts and the structure and the aspirations of LEED dovetailed perfectly with the aspirations CalPERS had set for their building." Lynn Simon, AIA, the project's green building consultant, says LEED established a green framework for the project and clarified the green building responsibilities of various team members. Perhaps more importantly, she remembers, it also "demystified green building" for many people involved with the project.

SITE PLAN

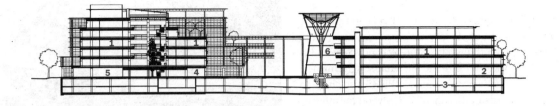

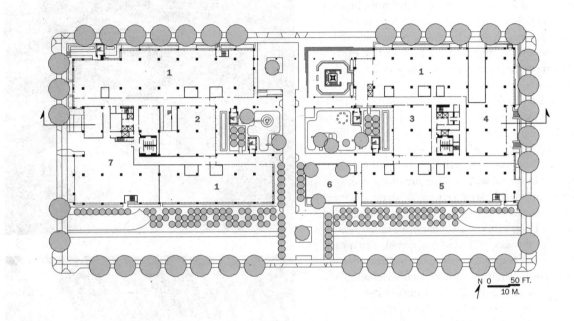

N 0   50 FT.
   10 M.

allowed designers the freedom to sculpt the building.

« The project features daylighting, views, natural ventilation, and a series of elevated terraces, fostering employees' connection with the natural environment.

Although the project achieved a LEED Gold rating, CalPERS's commitment to green design went beyond the rating system. Even though furniture was not covered by LEED at that time, for example, the team sought furniture with low-chemical emissions. CalPERS also chose to use Forest Stewardship Certified (FSC)-wood for most interior uses despite its cost-based decision to use conventional wood formwork, rendering the project ineligible for LEED's FSC-wood credit. In the end, a credit interpretation ruling on an unrelated project allowed CalPERS to remove the formwork from its FSC calculations, securing the credit after all. "There was never a decision to get a point just to adhere to the scorecard," says Markese. "If you do that, you might miss an opportunity to think about the building in a more holistic way."

Among the project's frustrations was the state's bidding system, which prevented the team from sole-sourcing specific products. Carpet tiles proved particularly frustrating for the project's interior architects. Doug Bregenzer, of Interior Architects, says, "To get three manufacturers that had high-quality products with recycled content, and to get that in a similar price range, was extremely challenging." The bidding system also meant that the project employed many more contractors and subcontractors than it might have otherwise, complicating communication and LEED documentation.

The lag between the project's start in 1999 and completion in 2005 meant that some design strategies and product choices seemed outdated by the time CalPERS occupied the building. The ballasted white roof, for example, did not qualify for the LEED credit for heat-island mitigation. "Today there are more options," says Simon. Bregenzer wishes the team had pushed harder for waterless urinals, which were uncommon at the project's outset but have gained a foothold in recent years. Of the five LEED points for water efficiency, CalPERS achieved only one, for reducing irrigation by 50 percent.

As is common in green buildings, CalPERS' mechanical system has required "a variety of tweaking to get it right," says Proctor. "We've just been in a year, so we're getting our feet on the ground and getting all the kinks worked out." Involving the commissioning agent in the design process instead of waiting for construction might have helped with that process and other issues, notes Simon.

CalPERS employees appreciate the new building, reports Proctor, but not as much as one might expect. While the occupants enjoy its daylighting and air quality, most of them relocated from Lincoln Plaza, which also has good daylighting and air quality. "[The new headquarters] building is beautiful," says Proctor, who is among the project's occupants, "but the people who are here moved from a very beautiful building. If they'd come out of a typical building," she notes "they'd have been blown away." Since the expansion was completed, CalPERS has hired Simon to direct Lincoln Plaza's application for a LEED for Existing Buildings certification. «

« The central atrium, referred to as the "living room," is topped by a skylight and a scrim that filters light and reduces heat gain. Many of the surfaces in this space are FSC-certified American cherry.

« The light-colored steel structure produces dappled shading in the outdoor courtyards.

SKY CONDITIONS

Sacramento's intense summer sun and winter cloudiness create challenging conditions for managing daylight.

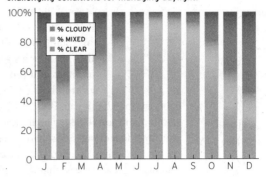

- % CLOUDY
- % MIXED
- % CLEAR

TEMPERATURES & DEW POINTS

The low dewpoint in summer suggests that sensible cooling, as opposed to dehumidification, is the primary challenge.

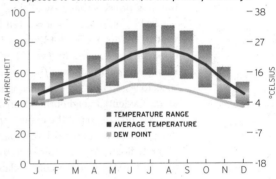

- TEMPERATURE RANGE
- AVERAGE TEMPERATURE
- DEW POINT

HEATING/COOLING DEGREE DAYS

Sacramento's relatively mild winters are reflected in the moderate number of heating degree days.

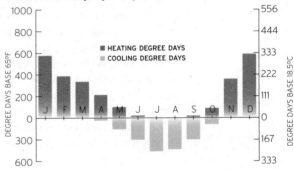

- HEATING DEGREE DAYS
- COOLING DEGREE DAYS

SECTION & PLAN DETAIL

1 Landscaped terrace/seating
2 Open office
3 High wall office/ vertical shaft
4 Corridor
5 Courtyard
6 Horizontal sunshade/integral planter
7 Horizontal sunshade/light shelf
8 Raised floor (HVAC/power/data)

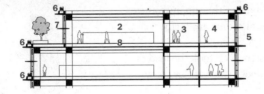

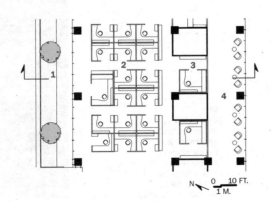

SOURCES

METAL/GLASS CURTAINWALL: Benson Industries

GLASS: Viracon (main building), Pilkington (entry pavilion)

CABINETWORK AND CUSTOM WOODWORK: Mid Canada Millwork

PANELING: The Freeman Corporation

WALLCOVERINGS: Carnegie Xorel

CARPET: Shaw carpet tile

INTERIOR AMBIENT LIGHTING: Zumtobel

KEY PARAMETERS

LOCATION:
Suitland,
Maryland
(Middle Potomac
watershed)

GROSS SQUARE
FOOTAGE: 208,000
ft²/19,320 m²

COST: $54 Million

COMPLETED: May
2006

ANNUAL ENERGY
USE (based on
simulation): 60
kBtu/ft² (690
MJ/m²)–19%
reduction from
base case

ANNUAL CARBON
FOOTPRINT
(predicted): 18 lbs.
CO_2/ft² (90 kg
CO_2/m²)

PROGRAM:
Offices, satellite
control rooms,
computer rooms,
conference
rooms, exercise
facility, cafe

NOAA TEAM

OWNER: General
Services
Administration

ARCHITECT:
Morphosis/
Einhorn Yaffee
Prescott

COMMISSIONING
AGENT: General
Services
Administration

ENGINEER: Einhorn
Yaffee Prescott
(MEP); Arup
(structural,
concept design);
Cagley and
Associates
(structural);
IBE Consulting
Engineers
(mechanical);
EYP Mission
Critical Facilities
(electrical); A.
Morton Thomas &
Associates (civil)

LANDSCAPE
ARCHITECT: EDAW

LIGHTING: Horton
Lees Brogden

ACOUSTICS: Shen
Milsom & Wilke

SECURITY: Jaycor

GENERAL
CONTRACTOR:
P.J. Dick

P HOTOGRAPHS FROM THE NATIONAL OCEANIC AND Atmospheric Administration's (NOAA) satellites— views of the earth and its weather patterns from above—pervade our media-rich culture. With this established public face, NOAA officials saw little need for a signature building by a big name architect. If it provided the tools they needed to control their satellites, a ho-hum building would do just fine, thank you. But the U.S. General Services Administration (GSA), which procures and manages facilities for most federal agencies, had other plans, and enrolled the project in its Design Excellence Program. One area in which the two agencies did agree was that the building should be green: GSA had just begun mandating a minimum LEED Silver rating when the project was announced, and NOAA sees itself as an environmental agency. "Our mission is environmental stewardship," says Paul Pegnato, NOAA's project manager for the facility. "Our building projects that stewardship."

Led by Thom Mayne (now a Pritzker Prize laureate), the joint venture of Morphosis and Einhorn Yaffee Prescott (EYP) developed a scheme based on several underlying goals. The first goal was to conceal as much of the building's required program space so as to lessen the visual impact of its volume on the site, which abuts residential neighborhoods in a Washington, D.C., suburb. The second was to put the majority of the employees on a single floor plate, to avoid the risk of departments getting broken up on separate floors. And the third was to provide an elegant, integrated solution that would accommodate the satellites, the people, and the technology that connects them.

These goals led to the unexpected design solution that features satellite dishes on the roof of a windowless rectangular box, dubbed "the bar," that houses the control rooms, while most of the employees work below grade in a cavernous, disk-shaped zone punctuated by light wells and skylights. A vegetated roof on the shallow dome over the main work space merges seamlessly into the landscape on the north, making most of the building's volume disappear from view. A glazed wall on the south, where the natural grade is lower due to the slope of the site, introduces light and views. Parking and mechanical rooms are farther below grade, underneath the main work space.

While Morphosis led the overall design, EYP took the lead on the green strategies. EYP project architect Doug Gehley (now

LEED SCORES
LEED-NC Version 2 Gold

	POINTS ACHIEVED	POSSIBLE POINTS
SITES [SS]	10	14
WATER [WE]	3	5
ENERGY [EA]	6	17
MATERIALS [MR]	5	13
INDOORS [EQ]	11	15
INNOVATION [ID]	5	5

MAXWELL MACKENZIE

» The satellite antennae and control rooms perch above an earth-sheltered office floor in NOAA's new operations facility.

Barometer of Change

with SmithGroup) says beyond the requirement for LEED Silver certification, little direction came from the agencies regarding environmental priorities. "The client left it wide open," he says. "Our goal was to sit in the meetings and watch for opportunities in the design as it started to develop." Once environmental opportunities were identified, teams that included designers from both Morphosis and EYP, and client representatives from NOAA and GSA, developed the solutions.

The mechanical engineers were charged with developing three designs for the building's systems. Of these, a system based on under-floor air delivery was chosen as the most effective way to provide comfort and fresh air to the occupants in a space with ceilings up to 28 feet high. Displacement ventilation with under-floor air provides other efficiency gains—including reducing the need to chill air for cooling—because it isn't being mixed with air that has already been in the space. Fan energy is reduced because the air is delivered at low speed and pressure. These benefits, combined with high-performance chillers and other measures, provide a predicted energy cost savings of 28 percent over the ASHRAE 90.1-1999 baseline. The under-floor air system also provides a level of individual control that would be tough for another system to match in an open floor plan.

Lighting the large, open work space was a challenge, according to Teal Brogden, senior principal at Horton Lees Brogden Lighting Design. One constraint was the mandate from GSA to have ambient lighting that provides at least 30 foot-candles of illumination on the work surfaces, even though individual task lighting was also available. "In some work situa-

The unexpected design features satellite dishes on the

tions we might take the ambient light levels down to 15 or 20 foot-candles," says Brogden. At the same time, the glazed wall on the south and the large light-wells create bright zones that had to be balanced to avoid uncomfortable contrasts. Based on lighting models, tubular skylights were added to enhance the daylight distribution, but "filling the space with the number of skylights that it would take to light with daylight was not in the original budget," notes Brogden. Instead, "daylight was meant to provide pools of visual interest and relief."

Another element that wasn't feasible because of lack of funding was operable shading devices on the vertical glass. Instead, the designers installed a fixed black scrim on the upper sections of the glazing to control glare, a solution that cuts down on the available daylight even when it is desired. GSA is considering removing that scrim, at least from the north and west sides, where it isn't needed to control direct sunlight, but no final decision has yet been reached. "We had suggested that they wait through the summer before they decide," says David Rindlaub, project architect with Morphosis. The risk of direct sunlight affecting workstations is mitigated somewhat by the high partitions in the systems furniture that GSA selected. While these partitions increase the amount of privacy in the individual cubicles, they also create a maze-like effect, and reduce the sense of spaciousness.

The combination of unusual form, high technology, and green measures made construction administration and commissioning a challenge.

SITE PLAN

1 Satellite operations
2 Green roof
3 Lower green
4 Parking

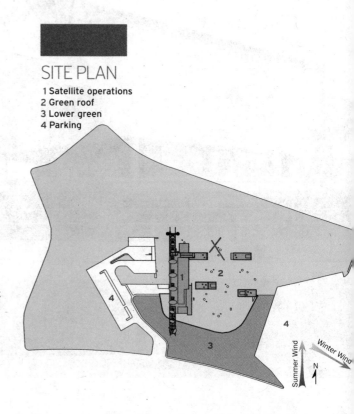

SECTION A-A

Please refer to the plan on the following page for section cuts.
1 Office 2 Exterior courtyard 3 Shared support ring 4 Parking

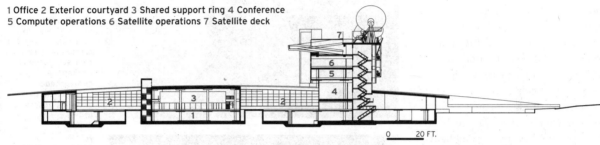

0   30 FT.

SECTION B-B

1 Office 2 Exterior courtyard 3 Shared support ring 4 Conference
5 Computer operations 6 Satellite operations 7 Satellite deck

0   20 FT.

roof—dubbed "the bar" of a windowless rectangular box.

≪
The entrance ramp
leads into a
conference facility
at an intermediate
level, below the
control rooms and
above the primary
workspace.

"There were some things that the contractor hadn't done before," notes Steve Baumgartner, who managed the commissioning process for EYP. In particular, keeping the under-floor plenum clean during construction was difficult. The technology challenge emerged when the engineers needed to commission the control room with actual electrical and thermal loads in place. "They wouldn't move the equipment in until it was tested, but we couldn't test without the loads in the spaces," says Baumgartner. Ultimately they found ways to simulate the loads that the equipment was expected to produce.

Yet another challenge on this project was the relatively high level of turnover among members of the design team. EYP's Gehley feels that this risk of turnover highlights the value of design firms that have enough depth in terms of green expertise and LEED-accredited professionals to carry a project forward when one person leaves.

While certification wasn't completed by press time, there is optimism that the project will exceed GSA's LEED Silver requirement and achieve Gold. It's too soon to judge how well the satellite control systems will work out, since the high-tech, mission-critical functions are still being fine-tuned. The design surely succeeds in fulfilling the goal of creating a provocative, iconic form. "Some people had a hard time, because it wasn't the conventional building they thought they were going to get," says Gehley. Among those who struggled through the process was NOAA's Pegnato, who still isn't convinced of the value of the building's drama. "Knowing what we know now, we could have tweaked the form to provide a bit higher level of function," he suggests. But he has no reservations about the sustainable agenda. "Relative to the green building, I would retain all aspects of the project." «

«
The extensive green roof is punctuated by large light-wells and circular skylights of various sizes.

»
Murals created using NOAA's own satellite images are printed on a scrim that separates the workspaces from an overhead walkway that provides access to utility spaces.

«
Ceiling heights up to 28 feet create a spacious feel in the large, earth-sheltered open workspace.

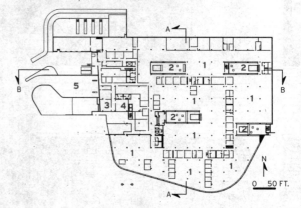

ENTRY LEVEL

1 Office
2 Exterior courtyard
3 Cafe
4 Gym
5 Ramp to parking

0   50 FT.

SKY CONDITIONS

It's challenging to control glare from direct sunlight on clear days without cutting off much of the natural daylight.

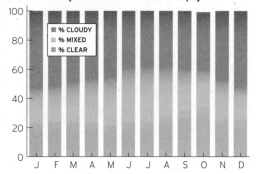

- % CLOUDY
- % MIXED
- % CLEAR

TEMPERATURES & DEW POINTS

The relatively high dew point (green line) reflects the region's high relative humidity.

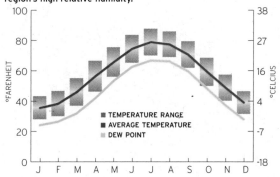

- TEMPERATURE RANGE
- AVERAGE TEMPERATURE
- DEW POINT

°FARENHEIT / °CELCIUS

HEATING/COOLING DEGREE DAYS

In spite of the relatively cool winter conditions, internal heat gains create a year-round demand for cooling.

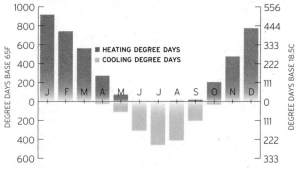

- HEATING DEGREE DAYS
- COOLING DEGREE DAYS

DEGREE DAYS BASE 65F / DEGREE DAYS BASE 18.5C

FOURTH LEVEL

6 Launch control
7 Satellite operations
8 Green roof

SOURCES

METAL/GLASS CURTAINWALL: PPG Sungate 100 low-e clear insulating glass

CLADDING: Swisspearl Carat, open joint fiber cement board panel system over Bakor Air-Bloc 33 vapor-permeable air barrier

DOWNLIGHTS: Focal Point Groove compact fluorescent with integrated custom acoustical panel. Acoustical panel uses Echo Eliminator Bonded Acoustical Pad

LOW-SLOPE ROOFING: Bakor 790-11 hot-fluid-applied rubberized asphalt membrane with 25% post-consumer recycled content is the waterproofing layer for green roof and non-green roof areas.

CARPET: Milliken Raffia Tex carpet tiles

Chapter 5

OFFICES

RESEARCH UNDERTAKEN BY CARNEGIE MELLON UNIVERSITY'S CENTER FOR Building Performance and Diagnostics, the Rocky Mountain Institute, and any number of government agencies has consistently shown that sustainably designed buildings lead to increases in worker productivity, decreased absenteeism, and a greater worker satisfaction rate. Design professionals have known this all along—after all, who wouldn't want to work in a daylit office designed around personal comfort, with no harsh chemicals and materials in the furnishings, and access to views and fresh air? But architects, engineers, and contractors have lacked the data to support using these sustainable design strategies. Furthermore, with few built examples to point out, they haven't been able to prove to clients that it was in their best economic interest to build a sustainable building.

Of course, the game has changed. Now, many clients are making demands of their design teams for supersustainable buildings, such as the five case studies presented in this chapter, routinely pushing the edge of contemporary green design. The National Association of Realtors' Washington, D.C., headquarters acts as a very public beacon of high-performance design to the real estate community. It's also a speculative office building, conforming to market demands and incorporating sophisticated items like daylight-dimming systems and variable-speed drives on mechanical equipment. Save the Bay, a non-profit organization in Providence, Rhode Island, saw its new office building as an opportunity to restore an environmentally dilapidated piece of riverfront property, a gesture that also fed its mission of preserving the nearby Narragansett Bay.

When it opened in 2004, the Alberici Redevelopment Corporation's light-filled office building in Overland, Missouri, was the highest-rated LEED Platinum project in the country with 60 points. The contractors for the non-profit Winrock International's new offices in Little Rock, Arkansas, didn't let the lack of available building-material recycling companies in town stop them from starting their own, resulting in 75 percent of the construction waste being diverted from landfills. Winrock was joined in Little Rock two years later by the new offices of the non-profit Heifer International, which through an ambitious daylighting strategy managed to reduce its energy needs by 55 percent in a conventional building designed to meet ASHRAE 90.1 standards. Skeptics of green building may point out that these case studies were built for non-profits, a trade association, and a corporation's headquarters—all of which are clients who tend to take a longer view with building projects, conceivably making sustainable choices easier on the design team. However, the truth is that each project had tight budgets, fast deadlines, and any number of roadblocks keeping them from achieving sustainable success. We would like to think the people who work in them would agree it was worth it. ◂◂

HE GREEN AGENDA WAS NEVER AN AFTERTHOUGHT for the National Association of Realtors' (NAR) Washington, D.C. offices—it was there from the beginning. But it was not the project's primary goal: Its aim was to create a landmark building that would give NAR a higher profile and better visibility (literally) from the halls of Congress. Gund Partnership and its consultants created such a landmark. "The taxi drivers all know it," says NAR spokesman Lucien Salvant. By using the LEED Rating System as a framework for design, the team also created a building that makes a strong environmental statement.

NAR searched all over the Capitol district for a location for its Washington, D.C., offices before settling on a small, derelict brownfield site. The project was built by local developer Lawrence Brandt as a build-to-suit speculative office building, of which NAR was to take ownership and occupy several floors. Special District of Columbia bonds, made available because the site was in a designated Enterprise Zone for economic development, helped finance the project.

Recognizing the triangular site's potential as a focal point—it is created by the intersection of New Jersey Avenue and First Street—NAR instigated a design competition to create the landmark building. Boston-based Gund Partnership won that competition and embarked on an odyssey that resulted in a building that is not only visually dramatic, but also Washington's first LEED-certified new building.

CASE STUDY
NATIONAL ASSOCIATION OF REALTORS BUILDING WASHINGTON, D.C.

NADAV MALIN

A GROUP THAT REPRESENTS THOUSANDS OF PEOPLE WHO PROFOUNDLY INFLUENCE THE CONSTRUCTION MARKET TAKES A STAND FOR THE ENVIRONMENT

A Beacon for Sustainability

KEY PARAMETERS

LOCATION: Washington, D.C. (Chesapeake Basin watershed)

GROSS SQUARE FOOTAGE: 100,000 ft² (9,300 m²)

COMPLETED: March 2004

COST: $28 million (base building only), $46 million total with NAR fit-out of 5 floors

ANNUAL ENERGY USE (BASED ON SIMULATION): 51,300 kBtu/ft² (583 MJ/m²)–39% reduction from base case.

ANNUAL CARBON FOOTPRINT (PREDICTED): 19 lbs. CO_2/ft² (94 kg CO_2/m²)–reduced by 50% during first two years by purchase of electricity from wind.

PROGRAM: Offices, retail

NAR TEAM

OWNER: National Association of Realtors

ARCHITECT AND INTERIOR DESIGN: GUND Partnership

ARCHITECT OF RECORD: SMB Architects

ENERGY CONSULTANT: Econergy International Corporation

COMMISSIONING AGENT: Advanced Building Performance

ENGINEER: CAS Engineering (civil); Fernandez & Associates (structural); E. K. Fox & Associates (MEP)

LANDSCAPE ARCHITECT: Oehme Van Sweden & Associates

LEED CONSULTANT: GreenShape LLC

LIGHTING: George Sexton Associates

AV CONSULTANT: Polysonics

FURNISHING CONSULTANT: Lucas Stefura Interiors

GENERAL CONTRACTOR/DEVELOPER: Lawrence N. Brandt

LEED RESULTS
LEED-NC Version 2 Silver

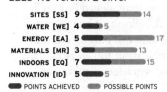

	Points Achieved	Possible Points
SITES [SS]	9	14
WATER [WE]	4	5
ENERGY [EA]	5	17
MATERIALS [MR]	3	13
INDOORS [EQ]	7	15
INNOVATION [ID]	5	5

POINTS ACHIEVED   POSSIBLE POINTS

«
The floor plate was expanded to a more economically viable size by obtaining a zoning ruling that allowed the designers to consider the entire facade as a bay window, meaning that it could protrude 4 feet (1.2 meters) beyond the property line.

» Retail and dining at street level are revitalizing the public space in the narrow, triangular block.

In addition to the unusual geometry of the site, the designers had to contend with a previously approved conceptual design that looked a lot like its neighbors, with concrete or masonry facades and punched openings. "These are very important places, these slim lots where roads come together at an angle," notes partner-in-charge Graham Gund, FAIA. "We felt that it called for more of an object building rather than being part of a continuous row."

Gund Partnership had a challenging community approval process to go through, while introducing their dramatic new design to replace the prior predictable one. After numerous meetings and hearings, the team ultimately prevailed after bringing a model of the neighborhood—large enough that it almost didn't make it through the door—to a final hearing.

Design and construction of the building was on a fast track: It took only two-and-a-half years from the conclusion of the design competition to NAR's occupancy of its space in October 2004 (and Gund Partnership was still finishing construction documents when construction began). In spite of the short-time frame, a number of innovative approaches for conditioning the space were explored. The tight floor-to-floor heights prevented

Partner-in-charge Graham Gund, FAIA, said "We felt that it called for more of an object building rather than being part of a continuous row."

chilled beams and under-floor air distribution; a double curtainwall would have taken too much valuable floor space, according to Laura Cabo, AIA, principal in charge. It helped that an adjacent building shades much of the west facade, reducing the need for extreme solar-control measures. Ultimately, low-emissivity double-glazing with a high-shading coefficient was deemed adequate for the curtainwall.

Interior shades are controlled manually to minimize glare and excessive energy use. Maintenance staff pull the shades down each night, then during the day employees adjust them for their own views and comfort. The interior color palette features warm, earthy colors to enhance the relaxing effect of the daylight, while the shades minimize lighting energy use.

Having been contaminated by past uses, including as a gas station, the site was an officially designated brownfield. As a result, before construction could begin, the contractors removed a 24-foot layer of soil and 533 gallons of groundwater. Given the speed of construction, reviewing submittals was a challenge. The LEED point for low-emitting adhesives and sealants was lost when a subcontractor unknowingly used the wrong product. And the team learned the hard way that the same carpet product can be procured with different levels of recycled content, depending on its exact construction. A product that they specified, believing it had 20 percent recycled content, arrived in a configuration that had only 5 percent.

NAR occupies five of the building's 12 floors, with about 100 employees in the association's government affairs, research, and regulatory departments.

« The glass facade offers a transparency not found in the neighborhood's more massive stalwarts.

SITE PLAN

1 Drop-off 2 Cafe and terrace 3 Water feature 4 National Park Service reservation

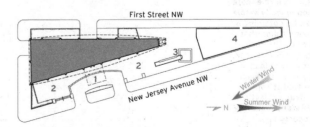

SECTION A-A

1. Recycled granite pavers
2. 10,000 gallon cistern for rainwater collection
3. Bicycle racks and designated alternative fuel vehicle parking
4. Photosensor & occupancy sensor for energy efficiency
5. Trees to reduce heat-island effect

6. Perimeter raceway for flexibility of space configurations
7. Shades for sun control
8. Perimeter diffuser for thermal control
9. Low-e glass to reduce heat gain and provide optimum views
10. Acoustic tiles with high recycled content

11. Placement of light fixtures to avoid light pollution
12. Carpet tiles with recycled content & low VOC limit
13. Waterless urinals & water efficient fixtures
14. HVAC with energy management control and CO_2 sensor control systems

15. 90% of habitable spaces have direct daylight and views
16. Drought-tolerant plants on terrace
17. Trellis for sun control at roof terrace
18. Light-colored membrane & pavers

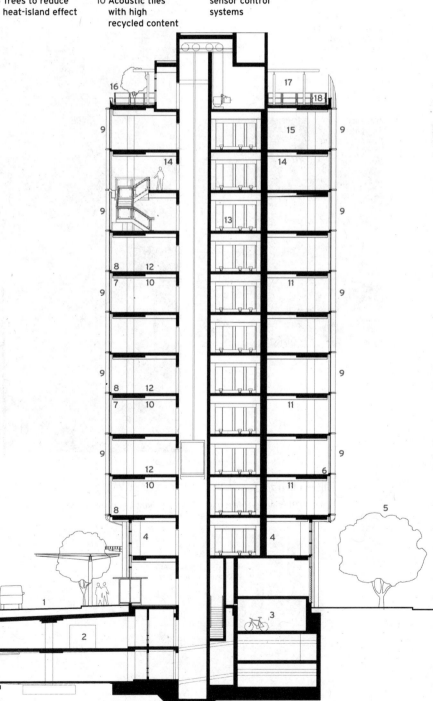

Fixed shading trellises on the south facade help manage solar gain.

The curtainwall glazing was specified to block nearly 90 percent of the solar heat, while still allowing significant visible light to permeate the space.

Light-colored pavers on the roof deck helped achieve the LEED point for urban heat-island reduction; rainwater drains in between the pavers and through pipes into the basement cistern.

HEATING/COOLING DEGREE DAYS

While heating degree days exceed cooling, lights, equipment, and people provide some "free" internal heat gain.

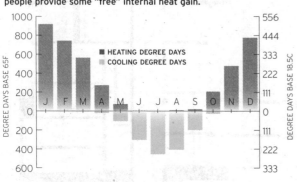

HEATING DEGREE DAYS
COOLING DEGREE DAYS

TEMPERATURES & DEW POINTS

The blue bars show normal monthly high and low temperatures over 30 years; the relatively high dew point reflects the high humidity.

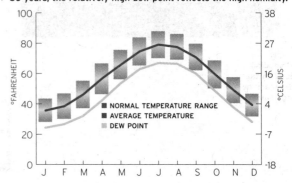

NORMAL TEMPERATURE RANGE
AVERAGE TEMPERATURE
DEW POINT

ROOF LEVEL

1 Terrace 2 Core mechanical

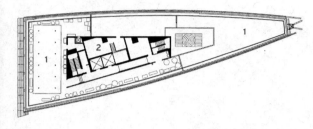

TYPICAL FLOOR

1 Circulation 2 Core 3 Open offices
4 Conference rooms 5 Private offices

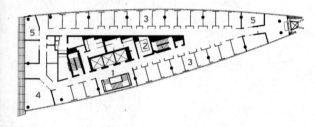

ENTRANCE LEVEL

1 Lobby 2 Retail 3 Garage 4 Service

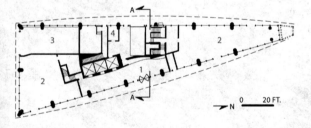

SKY CONDITIONS

The number of cloudy, mixed, and clear days in each month
changes relatively little over the course of a year.

NAR is fully occupied as of January 2008 and both ground floor restaurants are up and running. Tenants have aligned themselves with the sustainable aspects of the building and are using it as a mission-driven expression of their organizations. This includes the California Water Commission and the Concrete Institute, among others, each of whom have built out spaces according to LEED CI specifications.

As tenants change and floors are remodeled, the commissioning process continues, according to building manager Bradley Clark, of Cassidy & Pinkard. "Whatever results you had in the base building commissioning, you can throw them out when you do the tenant fit-out," Clark says. He was pleased to discover that the high-efficiency filters, used in the air-handlers to meet a LEED requirement, actually last twice as long as the simple glass mesh filters he used previously. Even though the new filters cost a lot more, he buys fewer and spends less time changing them, so in the end they provide cleaner air at no extra cost.

Clark has also found the waterless urinals "surprisingly easy to maintain." They work well, he says, and there is no odor except when changing the cartridges. He has found it necessary to change them religiously every two months, however, because "the moment the tenant realizes that they're bad, it's too late." As a result, NAR goes through a lot of cartridges, making these urinals more expensive to maintain than standard ones.

Another lesson came from the birds: Migrating swallows found the fixed louvers on the south facade to be an attractive perch, which meant that the windows had to be cleaned weekly. Once those louvers were outfitted with electrified wires, the birds were no longer a problem.

It has been Clark's experience that, in terms of energy performance, the building is not benefiting much from the variable-frequency drives (VFDs) on the air handlers because with small floor plates the demand for air is relatively uniform. Because "all the boxes are either open or closed," he notes, the reduced-airflow conditions under which the VFDs can save a lot of energy are uncommon.

Ambient lighting in the office space automatically dims when daylight is sufficient, so it is not unusual to see the lamps near the perimeter putting out much less light than those nearer the core. Employees have individual task lights as well; early on they tended to use the task lighting a lot, because the space they moved from was overlit and, by contrast, the new offices seemed dark. After a few days, however, the staff came to appreciate the calming effect of the lower-light levels, not to mention the great views, and began using the task lights less frequently. ◄◄

SOURCES

METAL/GLASS CURTAINWALL: Antamex International

WINDOWS: Viracon, VRE 7-38 (Alure) Front; predominant application; Viracon Low-E Glass VRE 3-38 (Gray) Side

CABINETWORK AND CUSTOM WOODWORK: PrimeBoard, Chesapeake Plywood

PAINTS AND STAINS: Sherwin-Williams, ProMar 400 (low-VOC paint)

FLOORING: Crossville: EcoCycle

CARPET: Interface Carpet Tiles, Kamala & Rice Wine

OFFICE FURNITURE: Herman Miller

LIGHTING: Alera Lighting and Edison Price Lighting

PLUMBING: Falcon Waterfree Urinals

CEILING: Armstrong, Interlude – Ceiling grids; Ultima Ceiling panels

SHADES: MechoShades

KEY PARAMETERS

LOCATION:
Providence,
Rhode Island (on
Narragansett
Bay)

GROSS SQUARE
FOOTAGE: 15,000
ft² (1,390 m²)

COST: $5 million
(building); $2
million (site and
infrastructure)

COMPLETED: July
2005

ANNUAL
PURCHASED
ENERGY USE
(BASED
ON
SIMULATION): 64
kBtu/ft² (730
MJ/m²), 32%
reduction from
base case

ANNUAL CARBON
FOOTPRINT
(PREDICTED): 11 lbs.
CO_2/ft² (54 kg
CO_2/m²)

PROGRAM: Offices,
classrooms,
meeting rooms,
dock, and
boathouse

TEAM

OWNER: Save The
Bay, Inc.

ARCHITECT/
INTERIOR
DESIGNER: Croxton
Collaborative
Architects

LANDSCAPE:
Andropogon
Associates

ENGINEERS: Lehr
Consultants
International
(mechanical/elec-
trical/plumbing/life
safety); Yoder +
Tidwell
(structural);
Northeast
Engineers (civil)

LIGHTING: William
Armstrong
Lighting Design

GREEN ROOF
CONSULTANT:
Robert Herman

ENERGY MODELING:
Quest Energy
Group

GENERAL
CONTRACTOR:
Agostini
Construction

Coast Guard

A NON-PROFIT'S REDEVELOPMENT OF A FORMER LANDFILL
SERVES AS A MODEL OF URBAN WATERFRONT RESTORATION

RUGGERO VANNI

SAVE THE BAY CENTER
PROVIDENCE, RHODE ISLAND

JOANN GONCHAR, AIA

Save the Bay first conceived of the idea of moving from its offices in a dilapidated former bank in the Smith Hill section of Providence in the late 1990s. The location, within striking distance of the Rhode Island State House, was well suited for the non-profit organization's lobbying and advocacy activities centered around restoring and protecting the Narragansett Bay and its watershed. However, it did not provide access to the waterfront needed for an expanding roster of educational programs, which included boat tours of the bay for children and adults.

Staff began searching along the industrial northern portion of the bay for a site suited for a boathouse and dock, as well as offices, classrooms, and meeting space. They identified a 6-acre waterfront parcel known as Fields Point, ideally situated near the head of the bay and just south of downtown. The group persuaded the owner, Johnson & Wales University, to donate the land, which in the not-so-distant past had served as a municipal landfill.

The selection of the former dump meant that design of the new Save the Bay Center would be just one part of a larger effort. "Our [first priority] was not to construct a green building, but to build on an urban brownfield on the coastline, in a bay-friendly, restorative way," says Curt Spalding, Save the Bay executive director. Even the building's architect agrees. "The coastal restoration dimension of the project was preeminent," says New York-based Randy Croxton, FAIA, principal of Croxton Collaborative.

To support the client's larger mission and meet programmatic needs, Croxton and its consultants wrestled with several basic issues. They asked themselves where to place the building to provide protection from storms and flooding, offer physical and visual access to the water, while not impeding development of habitat for marine life and other animals. They struggled with how best to minimize the building's footprint and its impact on the site. And they grappled with a tight budget of $7 million, with about $2 million of that total needed for extensive site work, such as capping contaminated soil, installing a methane venting system, and landscaping.

The scheme that emerged from this process is simple, yet striking. Completed in 2005, the 15,000-square-foot, one-story building has a vaguely V-shaped plan that follows the outline of Fields Point, and is situated about 65 feet from the shore. Two wings—one for classrooms and another for the office area—are "hinged" by a lobby that

frames a view of the bay. The steel structure, clad in wood clapboard and split-faced block, is sheltered with a series of stepped shed roofs partially covered with vegetation. From the vantage of the parking area, the green roof, along with a planted berm, creates the impression that the building has been "slid into the site," says Croxton. Inside, expanses of glass open it up to the water.

» A pergola shelters the lobby's glazing (opposite). Sail cloth baffles (top) help eliminate glare. From the parking area the building appears slipped into the site (bottom).

The illumination of the interior largely depends on daylight provided through skylights and clerestories. The wings' roughly east-west orientation permits use of simple strategies, such as extended overhangs to shelter the copious south-facing glazing from summer sun and prevent heat gain. Sail-cloth baffles suspended below the clerestories bounce light off the ceiling, allowing it to penetrate deeper into the space while eliminating glare.

A 4-ton, gas-fired heater/chiller provides space heating and cooling. Unlike conventional chillers, the system works without ozone-depleting refrigerants. Because it is fueled by natural gas, the equipment can reduce peak electricity load requirements. And it is capable of providing heating and cooling simultaneously to different parts of the building, explains mechanical engineer Val Lehr, PE, of the New York-based Lehr Consultants.

This dual operation is possible because the mechanical system includes eight air-handling units—a relatively large number for such a modestly sized building. The zones associated with these units are tied to occupancy sensors and the building management system, which can be set to allow temperatures to drift above or below the typical comfort range if spaces are not in use.

The building includes many other green strategies, including on-site power generation with a rooftop 20 kW photovoltaic array paid for by a grant from the local utility. It incorporates water-saving plumbing fixtures such as dual-flush toilets and waterless urinals. Cellulose wall insulation, paints without volatile organic compounds, and finishes with a high percentage of recycled content round out the more standard sustainable design choices.

In spite of all of the high-performance features, Save the Bay opted not to pursue a LEED rating. The decision was motivated in part by worries about costs that might be incurred for documentation, says Spalding. However, his main concern was that certification would require staff to redirect its attention from the organization's core activities of education and advocacy. "We spoke with other non-profits that had been through the process and LEED seemed to have consumed all of their energies," he says.

One of the costs that Save the Bay avoided was the expense associated with fulfilling the LEED commissioning prerequisite. However, commissioning might have prevented problems with the cooling system that it is now trying to resolve through retro-commissioning, says Omay Elphick, Save the Bay's on-staff project manager

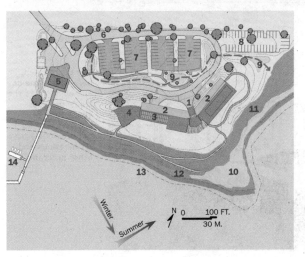

SITE PLAN

1 Arrival
2 Green roof
3 Photovoltaic array
4 Amphitheater
5 Boathouse
6 Bus parking
7 Parking
8 Oveflow parking
9 Bio-remediation swales
10 Restored marsh
11 Landscaped coastal buffer
12 High tide
13 Low tide
14 Dock

Winter
Summer

N   0   100 FT.
    30 M.

SKY CONDITIONS

The frequently cloudy conditions, especially in winter, make the abundance of daylight especially welcome.

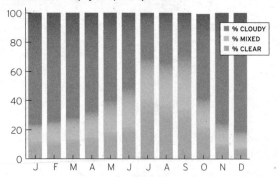

Legend:
- % CLOUDY
- % MIXED
- % CLEAR

TEMPERATURES & DEW POINTS

The relatively high dew point requires humidity control on incoming ventilation air.

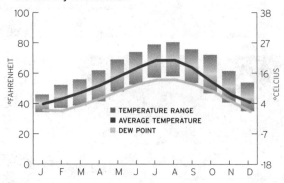

Legend:
- TEMPERATURE RANGE
- AVERAGE TEMPERATURE
- DEW POINT

HEATING/COOLING DEGREE DAYS

Both heating and cooling loads are substantial at this coastal New England location.

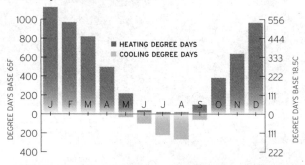

Legend:
- HEATING DEGREE DAYS
- COOLING DEGREE DAYS

⌃⌃ The office area with its many windows, clerestories, and baffle system, is largely dependent on daylight for illumination.

« The board room, overlooking the bay and adjacent to the lobby, provides space the organization did not previously have for meetings and other programs.

during construction. "In the category of lessons learned, commissioning is one of those things that can't be skimped on," says Elphick, now deputy director of Providence-based People's Power & Light, a buyer and seller of renewable energy contracts.

But despite some glitches, Elphick calls the project "precedent setting" as an example of urban waterfront redevelopment. Key to this redevelopment strategy was restoration of the site's "ecosystem services," such as cleansing the water that runs into the bay and promoting habitat for fish and other wildlife, says landscape architect José Almiñana, a principal of Andropogon, Philadelphia.

The building's roof plays a central role in this rehabilitation. In addition to merging the structure with the landscape, its vegetated surface absorbs and retains rainwater, allowing much of it to evapotranspire, reducing the amount of runoff that could mix with road salts, oils, and other contaminants, and ultimately be washed into the bay. The roof is just one piece of a site-wide strategy for capturing, filtering, and retaining stormwater that includes parking lots with pervious paving, a series of planted swales and trenches, and a buffer zone of native shrubs and grasses between the building and the shore.

«
After contractors completed shaping the shoreline with a revetment to prevent erosion from wave action, volunteers helped plant a buffer zone and salt marsh.

«
Upon entering the sky-lit lobby, visitors are greeted with spectacular and unobstructed views and a map of Narragansett Bay and its watershed on the floor.

SOURCES

METAL/GLASS CURTAINWALL: Vistawall Architectural Products
WINDOWS: Pella SmartSash III Casement Windows
GLASS: Viracon Solarscreen 2000 VE-2M
LOW-SLOPE ROOFING: White TPO UNA-CLAD UC-4 Roofing System
VEGETATED SLOPED ROOFING: Sarnafil
INSULATION: Cocoon Cellulose; Dow Styrofoam Cavitymate
INTERIOR PAINTS: Pittsburgh Pure Performance
PANELING: Muraspec Wall Panel Fabric
FLOORING: Forbo Linoleum; Shaw Carpets
INTERIOR AMBIENT LIGHTING: Linear Lighting; Cooper
Downlights: National Lighting; Mercury Lighting
CEILING PANELS: Armstrong
EXTERIOR LIGHTING: Cooper
LIGHTING CONTROLS: Lutron
PHOTOVOLTAICS: Kyocera KC167G 20kW photovoltaic array
CHILLERS: Yazaki CH-K40 4-ton Gas-fired heater chiller

ENTRY LOBBY SECTION

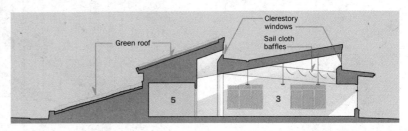

ADMINISTRATION WING SECTION

Along half its waterfront, the Save the Bay team created a salt marsh. Contractors removed fill and reshaped the land with riprap and clean sand. Then volunteers planted grasses, goldenrod, and elder. The intertidal zone provides an environment for shellfish and other marine life, prevents erosion, and traps sediment and pollutants.

Throughout the design and construction process, Save the Bay worked with Rhode Island's Coastal Resources Management Council to create policy that would encourage environmentally responsible redevelopment of contaminated waterfront properties and streamline the complex permitting process. New regulations were adopted in late 2006, and redevelopment of about 7,500 linear feet of urban shoreline along the northern Narragansett Bay and its tributaries is already under way. Known as the Urban Coastal Greenways policy, the regulations set standards for public access, stormwater discharge, and vegetation, and incorporate many of the restorative techniques employed at Save the Bay. That construction of his organization's headquarters would serve as a regional model seems to have taken Spalding by surprise: "We never thought our building would serve as a catalyst for restoration of the coastline." «

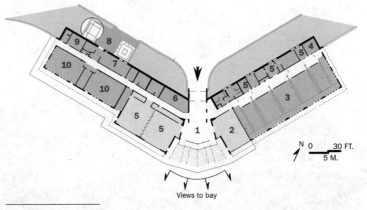

FLOOR PLAN

1 Lobby
2 Board room
3 Open office area
4 Lunch room
5 Conference/meeting rooms
6 Exhibition area
7 Mechanical room
8 Loading dock
9 Habitat lab
10 Classroom

The team forsook the original plan for a glass curtainwall in favor of a less expensive window wall. These cost savings went into higher quality, tinted, low-emissivity glass, which in turn allowed the company to forego window treatments, and save more money.

ALBERICI CORPORATE HEADQUARTERS OVERLAND, MISSOURI

JESSICA BOEHLAND

Top of the Charts

W E WERE GOING INTO UNCHARTED territory, where anything was possible," says Thomas Taylor, general manager of Vertegy, a subsidiary of Alberici Corporation, describing how his team achieved LEED certification for Alberici Corporate Headquarters. "Nobody on our team had ever worked on a LEED building, so nobody knew it couldn't be done." Which shows the importance of a good attitude: The building achieved 60 points—the highest LEED rating ever—on a budget of $147-per-square-foot, not including land acquisition or parking. "Sometimes being dumb is not so bad," laughs John Alberici, chairman of the board.

After a long search for a new home, the project team settled on a 14-acre site in Overland, Mo., near St. Louis.

A warehouse with three 70-foot-wide and one 90-foot-wide clear-span bays, each more than 500-feet-long, was still on the site. "The architect and I stood in the corner of that dim, dingy building," recalls Alberici, "and we could see from edge to edge, and he said, 'You know what we could do with this?'" That architect was John Guenther, AIA, principal at Mackey Mitchell Associates in St. Louis, who describes the space as a cathedral of steel.

The team restored the site with more than six acres of native prairie. Retention ponds and a constructed wetland treat stormwater on-site, while rainwater collected from the garage roof is used in the building's cooling tower and sewage conveyance system. Along with water-efficient fixtures, this practice reduces the building's potable water use by 70 percent, saving 500,000 gallons each year.

» The team refashioned the building's two northern-most bays as structured parking and removed the roof deck between the garage and the office space, creating a courtyard that lets light and air into both structures.

KEY PARAMETERS

LOCATION: Overland, Missouri (Mississippi River watershed)

GROSS SQUARE FOOTAGE: 110,000 ft² (10,200 m²)

COST: $21 million

COMPLETED: December 2004

ANNUAL PURCHASED ENERGY USE (BASED ON SIMULATION): 31 kBtu/ft² (350 MJ/m2), 60% reduction from base case

ANNUAL CARBON FOOTPRINT: (predicted): 10 lbs. CO_2/ft² (47 kg CO_2/m²)

PROGRAM: Office, conference, warehouse, parking

TEAM

OWNER: Alberici Redevelopment Corporation

ARCHITECT: Mackey Mitchell Architects

ENGINEERS: Stock & Associates, Civil Engineer; Alper-Audi, Structural Engineer

COMMISSIONING AGENT: Lillie & Co.

LANDSCAPE: Missouri Botanical Gardens & Shaw Nature Reserve

ENVIRONMENTAL CONSULTANT: Vertegy, an Alberici Enterprise

GENERAL CONTRACTOR: Alberici Constructors, Inc.

DESIGN-BUILD MECHANICAL/PLUMBING: Corrigan Company

LEED SCORES
LEED-NC Version 2 Platinum

	POINTS ACHIEVED	POSSIBLE POINTS
SITES [SS]	12	14
WATER [WE]	5	5
ENERGY [EA]	16	17
MATERIALS [MR]	9	13
INDOORS [EQ]	13	15
INNOVATION [ID]	5	5

SOURCES

STOREFRONT:
Vistawall FG-3000

SKYLIGHTS:
Naturalite

WOOD DOORS:
VT Industries

ACOUSTICAL CEILINGS:
Armstrong

CUSTOM WOODWORK:
Dow BioProducts; Smith & Fong

PAINTS AND STAINS:
Pittsburgh Paints and ICI

FLOOR AND WALL TILE: Marmoleum Tile, Forbo Flooring; Cork, Wicanders; EcoCycle Tile, Crossville Ceramics.

RESILIENT FLOORING:
EcoSurfaces

CARPET: Interface

OFFICE FURNITURE:
Herman Miller

LIGHTING: Ledolite; Capri Lighting; Day-Brite

EXTERIOR LIGHTING:
Gardco Lighting; BEGA

CONTROLS:
Johnson Controls

PLUMBING: Caroma Walvit (dual-flush water closet); Toto (Ecopower sensor); Waterfree (urinal)

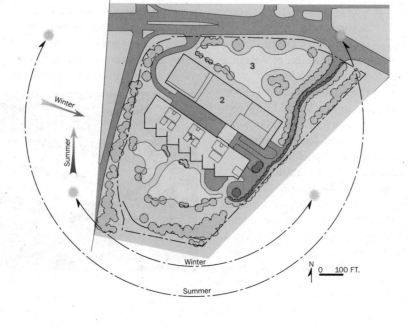

SITE DIAGRAM

1 New addition
2 Existing building
3 Pond

The building was designed to use 60 percent less energy than called for in ASHRAE 90.1-1999, and an on-site wind turbine and a solar water-heating system combine to meet 20 percent of its energy demand. A tight and well-insulated envelope, extensive daylighting, occupancy and daylight sensors on electric lighting, natural ventilation, heat-recovery mechanical ventilation, and a building automation system all contribute to the structure's low-energy demand. Adding offices in a sawtooth line along the building's southwest wall effectively reoriented the building due south, allowing it to take better advantage of daylighting.

When it first opened, the building was operating far less efficiently than it was designed to, but thanks to a sophisticated measurement and verification system, Taylor and his team brought performance in line with expectations, and they hope to push it further. In the latest round of operational changes, for example, the team widened the range of outside ambient temperatures for which the building management system activates free cooling. "What we found," says Taylor, "is that it's important to not just turn the keys over and walk away. You need to work with whoever's running the building so it can achieve its maximum performance."

SKY CONDITIONS

The mix of clear and cloudy conditions calls for a design that controls direct sunlight and provides daylight to the interior when it's cloudy.

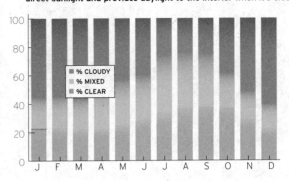

⋀
The addition of a sawtooth line of offices on the building's southwest wall created the daylighting benefits of a due-south orientation.

≪
While white predominates the interior for daylighting benefits, the design team used earth tones in strategic locations to bring more life to the space.

HEATING/COOLING DEGREE DAYS

Both winter heating loads and summer cooling loads are substantial.

TEMPERATURES & DEW POINTS

The combination of cold winters and hot, humid summers makes St. Louis a tough climate for low-energy building design.

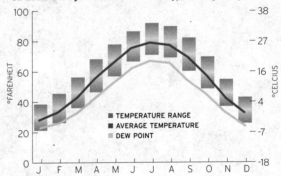

Alberici Corporation opted for an open interior to encourage interaction and camaraderie. (Alberici himself doesn't have a private office.) All occupied areas have daylighting, outside views, and operable windows. The company installed a white-noise system and sound-absorbent materials to preclude acoustics complaints common in open offices. Employees enjoy their workplace, and after the company's first year in its new headquarters, the human resources department reported a 50 percent reduction in sick days.

The building has also profoundly affected the project team members, who have since encouraged a market transformation in the St. Louis area. Alberici likes to say that at the beginning of the project he had a couple hundred green building skeptics, but at the end he had a couple hundred people looking for the next green project. ≪

Nonprofit's Mission Accomplished

WINROCK INTERNATIONAL EMBRACES
SUSTAINABLE DESIGN TO SUPPORT ITS MISSION
OF WORLDWIDE ECONOMIC DEVELOPMENT

Winrock International's new
headquarters is a testament
to the fact that green building
can be both attractive and
built on a conventional budget.

WINROCK INTERNATIONAL GLOBAL HEADQUARTERS LITTLE ROCK, ARKANSAS

NADAV MALIN

VERLOOKING THE ARKANSAS River in suburban Little Rock, Winrock International's new headquarters is "an abstraction of the traditional Arkansas dogtrot," according to design architect Kenneth Drucker, of HOK. Its signature feature, a gull-wing roof, not only keeps the sun at bay, it collects rainwater—an integrated solution that resulted from a collaborative and tightly managed design and construction process.

Winrock International is a nonprofit organization that supports sustainable development. In a staff of around 700 all over the world, about 70 employees are based at Winrock's headquarters in Arkansas, which was, until recently, on Petit Jean Mountain in rural Morrilton. That campus was expensive to maintain and required a lot of staff travel, so the board decided to find an alternative use for the facility and move its headquarters to Little Rock.

KEY PARAMETERS

LOCATION: Little Rock, Ark. (Arkansas River Valley watershed)

GROSS SQUARE FOOTAGE: 24,000 ft² (2,230 m²)

COMPLETED: December 2004

COST: $3.9 million

ANNUAL ENERGY USE (BASED ON SIMULATION): 29,600 kBtu/ft² (337 MJ/m²)–55% reduction from base case.

ANNUAL CARBON FOOTPRINT (PREDICTED): 14 lbs. CO_2/ft² (70 kg CO_2/m²)–reduced by 94% during first two years by purchase of electricity from wind.

PROGRAM: Offices and conference room.

WINROCK TEAM

OWNER: Winrock International

ARCHITECT AND INTERIOR DESIGNER: HOK

PROJECT MANAGEMENT: Horne Rose

COMMISSIONING AGENT: Cromwell Architects Engineers, Tao & Assoc.

ENGINEER: Cromwell Architects Engineers

LANDSCAPE ARCHITECT: Larson Burns & Smith

GENERAL CONTRACTOR: Nabholz Construction

LEED RESULTS
LEED-NC Version 2 Gold

SITES [SS]	6	14
WATER [WE]	4	5
ENERGY [EA]	9	17
MATERIALS [MR]	6	13
INDOORS [EQ]	13	15
INNOVATION [ID]	5	5

● POINTS ACHIEVED   ● POSSIBLE POINTS

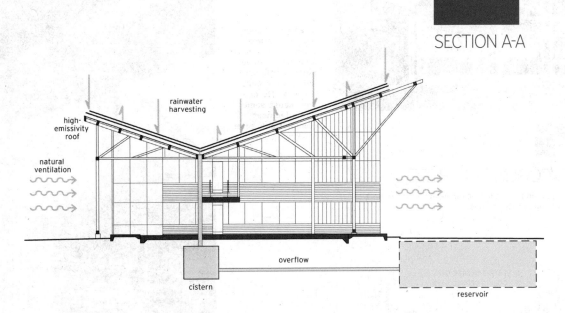

Labels in section diagram:
rainwater harvesting
high-emissivity roof
natural ventilation
overflow
cistern
reservoir

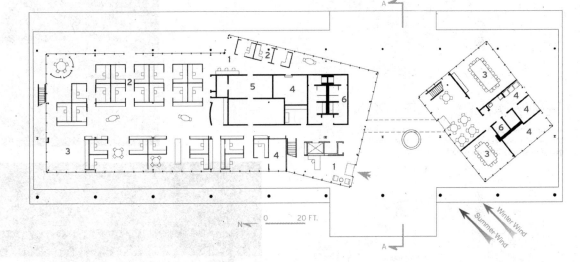

GROUND FLOOR

1 Reception
2 Offices
3 Conference
4 Utility
5 IT lab
6 Bath

N 0 20 FT.

A
A
Winter Wind
Summer Wind

« The expansive roof shades the building during working hours and collects more rainwater than the building needs for irrigation. Excess water is fed into the adjacent marina, replacing water that would otherwise be pumped from wells. The team was unable to get the city to approve the use of rainwater to flush toilets.

Given the organization's mission, making its new building green was a given. Winrock's president, Frank Tugwell, was familiar with green building, having worked previously on a project with green architect William McDonough. Winrock's board established two key goals for the project: that it would achieve a LEED Silver rating and that it would do that without costing any more than a comparable conventional building in Little Rock. The board set the budgetary constraint not only because of the organization's tight funds, but because they were determined to set an example for the local real estate market. "We wanted to be able to tell people who come in and are impressed, that it is something they can afford," says Tugwell.

Before hiring an architect, Winrock contracted with Horne Rose, one of several companies affiliated with green developer Jonathan Rose, to serve as the owner's representative. Horne Rose staff facilitated the process of selecting the design and construction team, then managed design and construction. Sarah Haga, of Horne

Rose, contends that conditions established at the inception of a project, such as how decisions will be made and what the schedule will be, can affect the outcome dramatically: "Our experience is that the best, most sustainable projects have a very thoughtful schedule that allows time to integrate green components," says Haga.

By showing examples of their previous work, principal-in-charge Bill Odell and his team from HOK convinced the board that they could produce a high-performance green building on a conventional budget. As soon as the architect and contractor had been selected, all the consultants participated with the client in a design charrette to develop the concept design.

The 2.2-acre site was tightly constrained by setbacks and other factors, leaving limited options for siting the building. There was only one way to orient the building in the available space, and it had the long axis running north-south. "We were not very happy when we saw the site," admits Odell.

After working through a series of shading studies, the

« Due to the building's suboptimal orientation with its long axis running north-south, sunlight on the walls in the late afternoon does cause some unwanted heat gain, but only after working hours.

HEATING/COOLING DEGREE DAYS

The demand for cooling in Little Rock is evident in the relatively high number of cooling degree days.

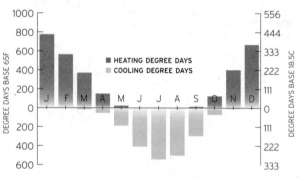

■ HEATING DEGREE DAYS
■ COOLING DEGREE DAYS

DEGREE DAYS BASE 65F: 1000, 800, 600, 400, 200, 0, 200, 400, 600
DEGREE DAYS BASE 18.5C: 556, 444, 333, 222, 111, 0, 111, 222, 333
Months: J F M A M J J A S O N D

TEMPERATURES & DEW POINTS

The blue bars show normal monthly high and low temperatures over thirty years; extremes on any given day may be much higher or lower.

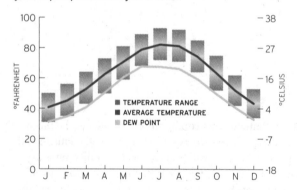

■ TEMPERATURE RANGE
■ AVERAGE TEMPERATURE
□ DEW POINT

°FAHRENHEIT: 100, 80, 60, 40, 20, 0
°CELSIUS: 38, 27, 16, 4, -7, -18
Months: J F M A M J J A S O N D

SKY CONDITIONS

The chart illustrates the average proportion of cloudy, mixed, and clear sky conditions over the course of a year.

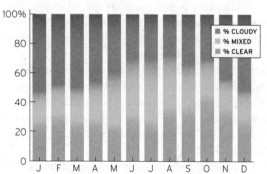

■ % CLOUDY
■ % MIXED
□ % CLEAR

Percent: 100%, 80, 60, 40, 20
Months: J F M A M J J A S O N D

« The offices and the conference room are separated by a breezeway that allows cooling air to flow, mimicking the traditional Arkansas dogtrot style.

team's solution to the less-than-optimal orientation was an oversize roof, which shades the building through most of its occupied hours. Some large trees were preserved nearby, helping to shade the building when the sun is low in the sky. The narrow plan that was dictated by the site's constraints helps with daylighting, nearly all of the regularly occupied spaces are fully lit during the day, so the lights don't need to be on.

Having the construction manager on board during design worked well. "We were able to run budgets on a number of design options through the course of design," says Stan Hobbs, AIA, of Nabholz Construction. "The whole experience was great for us," he adds, noting that it isn't always so smooth. Designers sometimes think "we're going to be telling them how to design the building," he says; on the contrary, Hobbs's goal is to help them figure out how to get their own design built.

Throughout the design process, solutions were proposed and then adjusted to meet budgetary constraints, which involved frequent negotiations between the contractor, the owner's rep, and the architects. "At first the entire building was glass," recalls Tugwell. When that turned out to be too expensive, some of the glazing was replaced by opaque siding. Nabholz was also helpful in

Texas, in 2002. Odell ran into two of them and was amazed at their enthusiasm. "They were working that expo. Not just picking up literature, but grilling the product reps.... That paid off with enormous dividends."

Prior to Winrock's construction, Little Rock had a minimal recycling infrastructure, so reaching the 50 percent recycling goal needed for achieving a LEED point seemed unlikely. But Nabholz staff catalyzed the creation of a local recycling system and achieved a recycling rate of 75 percent, which is worth two points in LEED, at no added cost. "That's something we now do on all of our projects, whether they are pursuing LEED or not," reports Hobbs.

Winrock's new headquarters has generated a lot of excitement in the community. Together with two other LEED buildings, the Clinton Presidential Library and the headquarters for Heifer Project International, this project has put Little Rock on the map as a green building tour destination. Tugwell and others frequently

SOURCES

METAL/GLASS
CURTAINWALL:
Guardian
Industries

WOOD DOORS AND
FLOORS: Algoma
Hardwood
Products and
K&M Bamboo
Products

DOORS: Algoma
Hardwoods and
Kawneer

ROOFING: Durolast

PAINTS AND STAINS:
Zolatone and
Sherwin-Williams

FLOOR AND WALL
TILE: Daltile

CARPET: Interface

FURNISHINGS: Knoll

LIGHTING:
Lightolier and
Ardon Mackie

PLUMBING: Zurn
and SCOT

Little Rock had a minimal recycling infrastructure, so reaching the 50 percent recycling goal needed for achieving a LEED point seemed unlikely.

«
With extensive
glazing and a
narrow plan, the
entire building is
bathed in daylight.

figuring out simpler ways to achieve the desired ends, which, Haga notes, saved energy during construction: "Constructability is looked at as a cost-saving measure, but it's also a sustainability measure."

The project's mechanical engineers were willing to explore unconventional approaches for providing comfort in a modern office building, but only up to a point. The entire office space has raised floors with under-floor air distribution, including a separate diffuser for each workstation so each occupant has control over temperature and airflow. But natural ventilation was not adopted, nor was the idea of taking advantage of the adjacent marina to use a water-source heat pump instead of a conventional cooling tower.

Although Nabholz Construction had not worked on a green building before, it took on the task with gusto. The firm sent several people, including the on-site supervisor for the Winrock project, to the U.S. Green Building Council's Greenbuild conference in Austin,

find themselves giving impromptu tours to visitors.

An extended process of shakedown and adjustments has followed initial occupancy, so the Winrock staff has yet to begin documenting energy performance. Among the significant changes was replacing the cooling tower with a heat-exchange coil in the marina. Fortunately, selling the cooling tower offset part of the cost of this modification. Additionally, lighting controls that were set up incorrectly are being rewired.

In the end, the most important test of a facility is whether the occupants like it. In this case, says Tugwell, there is no doubt that they do. "I think that they stay at work longer," he says. "They hang around because it is a pleasant place." «

Circle of Life

A CHARITY DEDICATED TO NOURISHING FAMILIES
BUILDS A NEW OFFICE AS A MODEL OF HARMONY WITH NATURE

TRISTAN KORTHALS ALTES

Built on a former Union Pacific railroad
switching yard, the Heifer International
Headquarters building's expansive views
extend beyond a set of railroad tracks to
the Arkansas River.

LEED SCORES
LEED-NC Version 2 Platinum

SITES [SS]	12	14
WATER [WE]	5	5
ENERGY [EA]	12	17
MATERIALS [MR]	6	13
INDOORS [EQ]	12	15
INNOVATION [ID]	5	5

POINTS ACHIEVED   POSSIBLE POINTS

TIMOTHY HURSLEY

EIFER INTERNATIONAL IS A NONPROFIT organization that addresses global problems with an approach founded in sustainability. It gives livestock such as goats, cows, and chickens to families in need as a lasting source of food and income. In 2000, during a period of strong growth and with its 200-plus staff spread across five locations in Little Rock, Ark., Heifer decided to build a headquarters to accommodate 450 (more than their current staff), and public space for educational and outreach programs.

Heifer moved into its 94,000-square-foot narrow, curving, four-story office building in January 2006, and its employees now work in daylit open offices with views of native landscaping, the Arkansas River, and the adjacent Clinton Presidential Library. The elegant and economical office building received the highest LEED rating—Platinum certification.

Reese Rowland, design principal with Polk Stanley, says sustainability was a top priority from the start. "Everything Heifer does is about sustainability," he says. While LEED certification was a goal, but "Heifer's interest was from an educational standpoint: We teach these values around the world; we need to demonstrate them in the way we live."

Delving into the organization and its history, the design team found a guiding metaphor in a statement attributed to Heifer founder Dan West: "In all my travels around the

KEY PARAMETERS

LOCATION: Little Rock, Arkansas (Arkansas River watershed)

GROSS SQUARE FOOTAGE: 94,000 ft^2 (8,730 m^2)

COST: $18.9 million

COMPLETED: January 2006

ANNUAL PURCHASED ENERGY USE (BASED ON SIMULATION): 33.6 kBtu/ft^2 (382 MJ/m^2), 43% reduction from base case

ANNUAL CARBON FOOTPRINT (PREDICTED): 12 lbs. CO_2/ft^2 (58 kg CO_2/m^2)

PROGRAM: Offices, conference, library, café, atrium

TEAM

OWNER: Heifer International

ARCHITECT/INTERIOR DESIGNER: Polk Stanley Rowland Curzon Porter Architects

LANDSCAPE: Larson Burns Smith

ENGINEERS: Cromwell Architects Engineers (MEP, structural); McClelland Engineers (civil)

COMMISSIONING AGENT: TME

ENVIRONMENTAL CONSULTANT: Elements

GENERAL CONTRACTOR: CDI Contractors

world, the important decisions were made where people sat in a circle, facing each other as equals." That sentiment is reflected in a set of concentric circles that create a sense of unity among the site's elements. Rippling outward from their center at a public entrance commons, the circles also illustrate the cycle of giving that Heifer calls "passing on the gift." The organization passes on charitable gifts in the form of livestock to needy communities; individuals or groups that receive the animals agree to share offspring with other community members.

» Exterior lightshelves, fins, and a wide roof protect the offices from the direct sun, while reflecting daylight deep into the interior.

Before work could begin on the design, the team had to deal with the 20-acre site Heifer had purchased, a former railroad switching yard that was polluted with creosote and diesel fuel. Instead of demolishing and trucking away several old warehouses that Heifer couldn't use, a subcontractor recommended a scheme that ultimately diverted 97 percent of the material from landfills. The refuse was processed on-site, with bricks set aside for reuse, rubble crushed for fill, and metals separated for recycling. The team had to decide whether to clean the contaminated soil or pay to have it taken to a landfill. Working closely with the city, they learned that the municipal landfill is required to bury each day's garbage with a layer of soil. The landfill took all 4,200 truckloads of Heifer's contaminated gravel, waiving the tipping fee, and used it instead of clean fill to bury garbage. This smart approach was "a benefit for both projects," says Dan Baranek, the project's civil engineer with McClelland Engineers.

The property's proximity to the Arkansas River made the team acutely aware of the consequences of runoff. "We set a goal early to manage stormwater on-site," says Rowland. Thus, the grounds collect and contain stormwater in a wetland (lined with clay excavated from the parking lot), which surrounds the building and snakes through the property.

A permeable paving system in the parking lot encourages stormwater infiltration, while a 3,000-gallon tower collects rainwater from the building's 30,000-square-foot roof. That water supplements a separate graywater storage tank fed from lavatories and condensate from ventilating units; together the systems supply water for toilets and the cooling tower, which account for 90 percent of the building's water needs. Under a special vari-

FIRST FLOOR/ SECTIONS

1 Entrance and bridge
2 Wetland
3 Atrium
4 Water tower
5 Open offices
6 Large meeting room
7 Outdoor meeting room
8 Mechanical

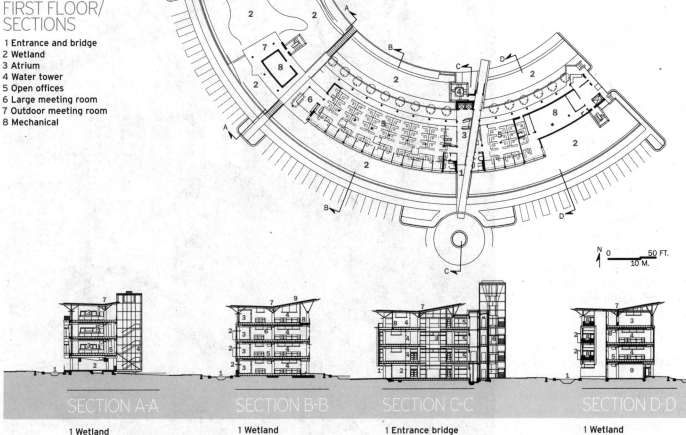

N 0   50 FT.
   10 M.

SECTION A-A

1 Wetland
2 Path to commons under building
3 Cafe/break area
4 Cantilevered stairwell
5 Typical raised floor
6 Typical return air
7 Water collection

SECTION B-B

1 Wetland
2 Lightshelves
3 Office
4 Open office
5 Typical raised floor
6 Typical return air
7 Water collection
8 Balcony
9 Future photovoltaics

SECTION C-C

1 Entrance bridge
2 Vestibule
3 Atrium
4 Conference
5 Elevator lobby
6 Water tower and stair
7 Water collection
8 Balcony

SECTION D-D

1 Wetland
2 Lightshelves
3 Office
4 Open office
5 Typical raised floor
6 Typical return air
7 Water collection
8 Balcony
9 Mechanical room

« All stormwater is treated on site. A sculpted wetland surrounds the building, providing a wildlife habitat in view of the offices. Cool air off the water helps condition the building's staircases.

« Open offices allow collaboration among staff. Sensors adjust the amount of electric light needed in response to daylight.

ance from the city, restrooms have
waterless urinals. Even the native-
plant landscaping only required irrigation to get established.

Although only 40 percent of the
headquarter's envelope is glass, the
interior feels bright, in part because
of the innovative placement of the
building's three staircases. Instead of
hiding them inside, they are individu-
ally articulated and wrapped in glass.
Two float over the wetland. With air
openings at ground level and five stories up, convection pulls cool air off
the water, helping to keep the uncon-
ditioned spaces comfortable year-
round. The third wraps around the
water tower. "The staircases promote
health and they cut down on elevator
use," says Rowland. Five balconies on
each floor create additional opportunities to step outside.

From the beginning, Rowland says,
they had hoped to use 35 percent less
energy than a standard ASHRAE
90.1 building, but modeling showed
they could save as much as 55 percent. "Optimized daylighting
throughout the building was the key
to that success," says Todd Kuhn, the
lead mechanical, electrical, and
plumbing engineer. Together with
photocells and dimming controls,
occupancy sensors reduce energy use
from lighting and lower the cooling
load. "Everyone loves the daylighting," says Erik Swindle, Heifer's director of facilities management. Despite
modeling during the design process,
some adjustments were needed, he
adds. "The architect had said we did-
n't need window shading," but shading was installed on the building's
south side shortly after moving in
because of the glare in some offices.

The commissioning process was
also essential to realizing projected
energy savings. A series of check
valves on the HVAC system was
removed after a value engineering
evaluation, for example. "We moved
in during the winter, and our heating
and cooling systems were competing
with each other," says Swindle. "We
could have spent $20,000 to
$30,000 to install those valves, or we

SITE PLAN

1 Entrance
2 Constructed
  wetland
3 Permeable
  parking
4 Typical bioswales
5 Water tower
6 Future wetland

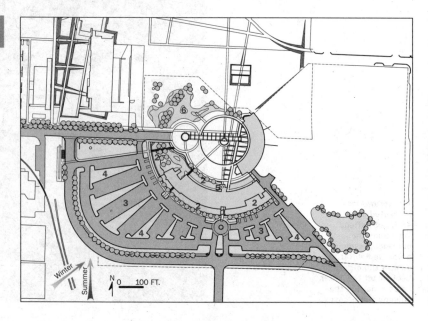

Rainwater from the building's 30,000-square-foot roof collects in a five-story, 3,000-gallon tank, shown at right. The building's toilets and cooling tower use that water, along with water from a graywater recycling system.

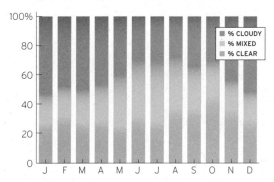

SKY CONDITIONS

A high incidence of cloudy conditions throughout the year drove the design team to maximize daylighting in the building.

- % CLOUDY
- % MIXED
- % CLEAR

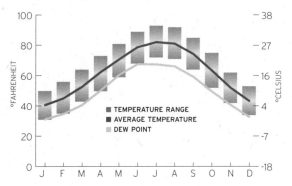

TEMPERATURES & DEW POINTS

High relative-humidity leads to condensate generated by the building cooling system, which is used to flush toilets.

- TEMPERATURE RANGE
- AVERAGE TEMPERATURE
- DEW POINT

°FAHRENHEIT / °CELSIUS

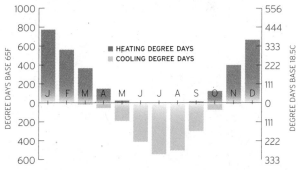

HEATING/COOLING DEGREE DAYS

Little Rock's climate necessitates a high amount of both heating and cooling throughout the year, calling for efficient HVAC systems.

- HEATING DEGREE DAYS
- COOLING DEGREE DAYS

DEGREE DAYS BASE 65F / DEGREE DAYS BASE 18.5C

could have wasted that much energy every year," he says, noting that the decision was easy. Another fix prescribed by commissioning helped tighten up the under-floor air distribution system. Leaks around data boxes in the access flooring unit that caused unnecessary noise and energy loss were sealed.

"Throughout the project we kept a pretty lengthy running shopping list of sustainable strategies," says Rowland. The team ended up checking off almost everything on that list, even when doing so cost them LEED points. Instead of buying certified wood from across the country, the team bought noncertified wood from a responsibly-managed forest in the region. This local approach paid off in other areas of LEED. For instance, the team found a high percentage of building materials within 500 miles of the site by buying steel and aluminum from manufacturing facilities in Little Rock.

The building was completed for $19 million, or a modest $189-per-square-foot—important for an organization supported by individual donors who demand fiscal responsibility. A future welcome pavilion will include galleries and a shop, and an interactive learning center will model village life around the world—with livestock, of course. The open quarters required an initial adjustment period for staff, yet "once we settled in, people started recognizing details," says Swindle. "Everybody loves watching the plants grow, and it was great to see ducks land in the water three weeks after we moved into the building." ≪

SOURCES

METAL/GLASS CURTAINWALL: Kawneer 1600 series

GLASS: Guardian 5G 5N-68; LOW-E

LOW-SLOPE ROOF: Genflex TPO

THERMAL INSULATION: BioBased

SOUND WALL INSULATION: Bonded Logic

CABINETWORK AND CUSTOM WOODWORK: Solid surfacing: Avorite-K3-8200, Kaleidoscope

TOILET PARTITIONS: Yemm & Hart

FLOORING: Hanlite Enterprises

CARPET: Interface

OFFICE FURNITURE: Steelcase (Kick)

TASK CHAIRS: Steelcase (Think)

INTERIOR AMBIENT LIGHTING: Corelite (Vertech pendent)

EXTERIOR LIGHTING: Invue: Strut

LIGHTING CONTROLS: Leviton (Centura daylight dimming system)

CHILLERS: Carrier

Chapter 6

RESIDENTIAL

FROM A LUXURIOUS $100 MILLION MANHATTAN HIGH-RISE TO A $22 MILLION shelter for adults in need of a home in San Francisco and an $11.2 million low-income residence in Chicago, the projects featured in this chapter aim to, at a minimum, provide comfortable accommodations, while also incorporating the latest in green building practices. Common tactics such as the use of low-VOC materials and solar photovoltaic arrays are employed in the buildings to minimize each project's carbon footprint, yet the buildings also exemplify distinctive solutions for sustainability rooted in the local site conditions. Perhaps most important, each project builds on an existing urban fabric by using brownfield sites and tapping into utility and transportation infrastructure already in place.

Near North, a 96-unit Chicago residence for low-income tenants, has a wind-turbine system on its roof to curb electricity costs. In the works is a water-recycling system that will collect drainage from sinks and showers, then filter the water for reuse in flush toilets. When the system is implemented, it is estimated that each year the project will save 45,000 gallons from being pumped from nearby Lake Michigan. San Francisco's Plaza Apartments house once-homeless adults in the city's skid row. The architects designed it to minimize solar gains, which can still be significant even in a city known for its foggy, temperate climate. A rain-screen cladding system allowing air to circulate behind the façade alleviates heat gain from sunlight, and the windows on the southwest façade are recessed to provide afternoon shade.

On Manhattan's West Side, tenants of the 38-story, glass-clad Helena share more than corridors and elevators. A common-water-source heat-pump system balances the heating needs from one apartment to another. Upon moving in, tenants of the building receive a welcome package that encourages them to use eco-friendly cleaning products. Also included is information about how to economically operate apartment appliances. This gesture is a reminder that, in addition to sustainable building practices, it is the small and simple efforts taken by individuals to consciously reduce their carbon footprints that help make an impact on reversing the environmental effects of urban living. «

Second Acts

MURPHY/JAHN RETHINKS LOW-INCOME
HOUSING FOR A REVITALIZED CABRINI-
GREEN IN CHICAGO

NEAR NORTH APARTMENTS
CHICAGO, ILLINOIS

DAVID SOKOL

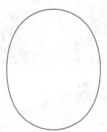

BSERVANT CHICAGOANS WILL note the aesthetic similarities between Near North Apartments, the Murphy/ Jahn-designed building in the city's Near North neighborhood that opened in March, and the architect's four-year-old student housing project at the Illinois Institute of Technology. Like State Street Village dormitory, Near North is a curvilinear wedge of a building—a giant bread loaf whose raked angles create a memorable presence along the streetfront. Both buildings have reinforced concrete structures with facades clad primarily in corrugated stainless steel and lined with punched windows to create what Murphy/Jahn principal architect Scott Pratt calls "a cellular expression of the building." And the dormitory rooms, laid out with en-suite bathrooms and the occasional kitchenette, approximate the apartment living at Near North.

The buildings' ambitions couldn't be more different. If State Street Village is a temporary perch for ambitious students on the road to success, the 96-unit Near North Apartments is a place for starting over. The single-room occupancy (SRO) is operated by Mercy Housing Lakefront and offers permanent supportive housing for disabled, formerly homeless, and Chicago Housing Authority residents. While a student may be

KEY PARAMETERS

LOCATION: Chicago, Illinois (Chicago River watershed)

GROSS SQUARE FOOTAGE: 45,810 ft^2 (4,260 m^2)

COST: $11.2 million

COMPLETED: March 2007

ANNUAL PURCHASED ENERGY USE (BASED ON SIMULATION): 56.2 kBtu/ft^2 (638 MJ/m^2)

ANNUAL CARBON FOOTPRINT (PREDICTED): 21 lbs. CO_2/ft^2 (100 kg CO_2/m^2)

PROGRAM: 96 units SRO, reception and lobby, community room, case management office, building management offices, support facilities (laundry, bike storage, tenant storage, etc)

TEAM

OWNER: Mercy Housing Lakefront

ARCHITECT: Murphy/Jahn

ASSOCIATE ARCHITECT: Smith and Smith Architects

LANDSCAPE: Terry Guen Design Associates

ENGINEERS: Graeff, Anhalt, Schloemer & Associates (civil/structural); Environmental Systems Design (MEP)

COMMISSIONING AGENT: The Renschler Company

GENERAL CONTRACTOR: Linn-Mathes

^^ A curved, reflective skin of steel and glass encloses the structure (top). On the roof, eight cylindrical Mylar-finned wind turbines take advantage of southwesterly winds, generating power to heat water and supplement the building's electrical needs. The roof's curves at the edges drives wind into the turbines. (above).

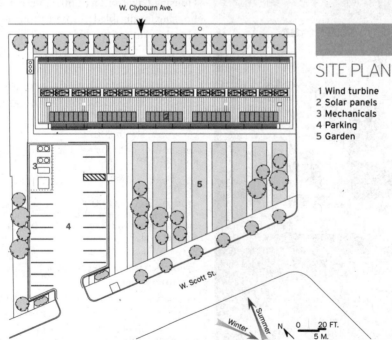

W. Clybourn Ave.

SITE PLAN

1 Wind turbine
2 Solar panels
3 Mechanicals
4 Parking
5 Garden

W. Scott St.

Winter   Summer   N   0   20 FT.
                          5 M.

SOURCES

GLASS CURTAINWALL: Traco TR-7801 Curtainwall

GLASS: Pilkington Solar E Glass Insulated Units, fabricated by J.E. Berkowitz

WIND TURBINES: Aerotecture International

SOLAR THERMAL PANELS: Solargenix

CHILLER: Carrier

BOILERS: Viessmann Vitodens 200 Condensing Boiler

GRAY WATER SYSTEM: Green Turtle Technologies

catching a movie in the communal lounge, 43-year-old Near North tenant Thomas Gooch visits his on-site case manager to discuss his recovery from substance abuse. Where the young guns crack the books in their rooms, people like 52-year-old Donnie Conner, who hopes to earn two associate degrees from local Harold Washington College, are living out a second chance.

Just as the members of the new community at Near North partake of very different privileges from the students at IIT, their buildings also differ in an important way. Unlike its predecessor, the SRO elegantly melds architectural form with an environmental agenda designed to achieve a LEED Silver rating.

Near North represents a do-over in other respects. The neighborhood had been home to Cabrini-Green, towers built and maintained by the housing authority and which exemplified everything that could go wrong with mid-century urban renewal. The city came to view the development's social stigma and high crime as unacceptable. The zoning code was rewritten to encourage dispersed low-income housing, and in 2000 Cabrini-Green was scheduled for demolition.

Barry Mullen, Mercy Housing Lakefront's vice president of real estate development, explains that Chicago Mayor Richard Daley launched a corresponding initiative to build new supportive housing. In 2003, an RFP was issued for 32,500 square feet of remedi-

ated brownfield located in the heart of Near North, and Lakefront Supportive Housing, which merged with Mercy Housing Midwest in January 2006, received authorization to develop the site. "The city basically makes a donation of the land," Mullen says, "and we go about putting the financing together. Our tenants pay a third of their income as rent, so these supportive-housing projects are almost always done with grants, subsidies, or tax-credit dollars."

Such minimal rental income also galvanized the client's move toward sustainability. "If we can reduce the operating cost on these buildings through green technology or any other tool, we will explore it and take it to its logical conclusion," Mullen says.

Indeed, the organization had implemented green-building strategies in previous projects like Wentworth Commons Apartments, which included rooftop photovoltaics and bioswales. For Near North, it would also incorporate great design—"to make a statement that housing for low-income people can strive for excellence as much as any other building," Pratt says. Mercy Housing Lakefront board member and local construction executive Harold Schiff recommended teaming up with Murphy/Jahn principal Hemut Jahn, who agreed to work for a reduced fee.

To achieve such excellence, "Helmut always says that when nothing can be added and nothing can be taken away, you've got the right design," Pratt says, pointing out that Near North's

SKY CONDITIONS

Chicago's moderate mix of cloudy, mixed, and clear skies is relatively consistent throughout the year.

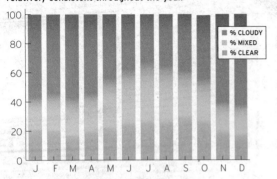

Legend:
- % CLOUDY
- % MIXED
- % CLEAR

TEMPERATURES & DEW POINTS

The heat of Chicago summers and cold of the winters suggests a contained building, easily sealed off from the elements.

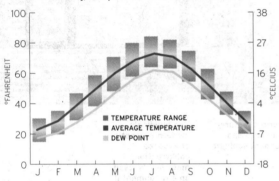

Legend:
- TEMPERATURE RANGE
- AVERAGE TEMPERATURE
- DEW POINT

HEATING/COOLING DEGREE DAYS

Heating loads dominate the comfort model, but cooling loads to provide comfortable indoor spaces, are not insignificant.

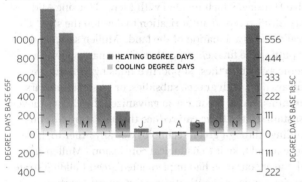

Legend:
- HEATING DEGREE DAYS
- COOLING DEGREE DAYS

The clients used the cachet of a Murphy/Jahn-designed structure to change the perception of SRO buildings.

bread-loaf form and sleek skin synthesized the client's architectural and sustainability goals better than alternative design concepts. For example, the stainless-steel wall and roof cladding has a high albedo, reducing the urban heat-island effect. The roof's slightly peaked shape and orientation to the northeast and southwest are perfectly suited for drainage into a 1,500-gallon rainwater cistern used for landscape irrigation, too, and also to accommodate the placement of city-donated solar thermal panels that supply 30 percent of the building's domestic and heating hot water.

The overall profile of the building also has green benefits. "You can see from fluid dynamic studies that airflow is very smooth over the center of the building," Pratt says. To exploit that potential, Mercy Housing and Lakefront, with the Murphy/Jahn team, reserved the roof's peak for an innovative horizontal wind-turbine system that currently meets 8 percent of the building's electricity demand.

The 520H Aeroturbine was invented by University of Illinois professor Bil Becker to generate power from the wind in urban settings. It sports a petite, modular design that can accommodate city rooftops of many sizes (Near North actually includes eight modules linked together). Functionally, the Aeroturbine features both Savonious and Darrieus rotors. These two different airfoils perform ideally at different wind speeds, so that the hybrid yields continuous energy production in variable city winds. Moreover, the helical Savonious does not rotate in excess of 400 revolutions per minute (rpm), which prevents electrical surges in high winds, ice throwing, or a level of vibration that could disturb tenants.

Although Becker has been working on the Aeroturbine and its sister, the vertical 510H, since earning a research grant from the Carter administration in 1979, the Near North installation represents only his second time installing a 520H system. Feeding directly to inverters, it also is the first battery-free wind turbine, he claims. "We have the opportunity to do things here that others might not," Mullen says of Mercy Housing Lakefront's willingness to try out the technology, which was paid for by a grant from Illinois Clean Energy Community Foundation. "We're happy to share all performance data. Somebody had to do it first."

Another first for Near North is its graywater system, which had not been attempted in the city. Despite Chicago's green reputation, Sadhu Johnston, Deputy Chief of Staff to the Mayor and former commissioner of the Department of Environment, confirms, "I don't think we're leaders in this technology. A lot of people would think, 'I don't need to save water—the Great Lakes have 20 percent of the world's fresh water.'" Drainage is collected from showers and sinks, and then filtered, treated, and reused to flush toilets. Project consultant Dan Murphy, PE, of Environmental Systems Design, says the measure should keep 45,000 gallons from being pumped from Lake Michigan every year.

SECTION A-A/SECTION DETAIL

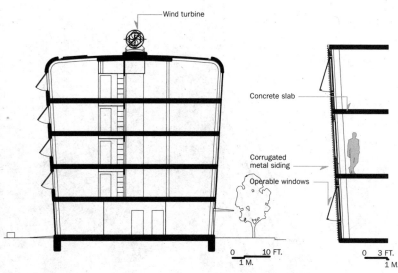

Wind turbine

Concrete slab

Corrugated metal siding

Operable windows

0   10 FT.
1 M.

0   3 FT.
1 M.

« The ground floor of the Near North Apartments features a community room that will be open to neighbors as well as residents.

TYPICAL FLOOR

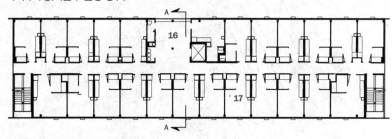

A

16

17

A

GROUND FLOOR

15   13   3   2

1

7   9   4   11

4

14

10   12   4   5   6   8   4   4   10

N   0   20 FT.
5 M.

1 Lobby
2 Vestibule
3 Reception
4 Office
5 Trash/recycling
6 Laundry
7 Smoking lounge
8 Pantry
9 Community room
10 Storage
11 Conference room
12 Bicycle room
13 Maintenance
14 Electrical closet
15 Mechanical room
16 Public area
17 Typical apartment

Unlike the rooftop wind turbines, Near North's graywater technology was greeted with less alacrity. While the city's Department of Construction and Permits breezed through approval with only a hint of caution, the Illinois Department of Public Health temporarily shut down the system earlier this spring. Johnston says the state's slower acceptance owes to the rarity of the technology's installation. He also notes, "We're figuring out how we want to work with graywater in the future. [Near North] raised a whole series of questions that we're developing answers to."

While the city prepares its guidelines for future graywater systems, Near North continues improving itself. Inside the building, a touchscreen educates residents about green living. Meanwhile, Becker has been tweaking his Aeroturbines, admitting that the system passes along about 60 percent of the energy it captures. Part of the inefficiency has to do with the wind interface units, which play electronic middleman between the turbine and the inverter: "Our machines don't spin up fast enough to create a voltage surge, but we've got all these capacitors and filters and buffers essentially braced to take it," he says of the off-the-shelf parts. Currently, he is reprogramming the interface units' maximum power point settings to complement Aeroturbines' behavior, and says to expect a jump in average production from 200 kWh per module per month to 300 kWh.

Becker adds that other elements of the system could be improved—he's still searching for an alternator optimized for lower rpm—but some of Chicago's decision-makers are convinced already. Johnston says that an Aeroturbine installation will be mounted atop a municipal high-rise in the near future. «

CASE STUDY
THE HELENA
NEW YORK CITY

JOANN GONCHAR, AIA

LEED SCORES
LEED-NC Version 2 Gold

SITES [SS] 10 ●━━━ 14
WATER [WE] 5 ●━ 5
ENERGY [EA] 6 ●━━━━ 17
MATERIALS [MR] 5 ●━━ 13
INDOORS [EQ] 10 ●━━━ 15
INNOVATION [ID] 5 ●━ 5
● POINTS ACHIEVED ○ POSSIBLE POINTS

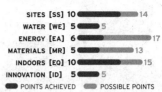

A Westside Story

THEY MIGHT BE CONSIDERED URBAN pioneers, but the approximately 1,400 tenants of the "Helena," one of New York City's first green high-rise residential buildings, are hardly roughing it. The 38-story, 580-unit building on the western edge of Midtown is not just basic shelter. Its features—Hudson River views, blonde wood kitchen cabinets, and an on-site health club—are the stuff of real-estate obsessed Manhattanites' dreams.

The Helena, which opened in late 2005, offers more than amenities, views, and nice finishes. Behind the $100 million tower's gleaming glass and metal skin is a serious high-performance building. Its owner and designers say the Helena will annually use 65 percent of the energy and one-third of the potable water of comparable properties.

These strategies help satisfy a growing demand in New York City's real estate market, according to the

Helena's owner. "We used to believe that green features helped a building lease up faster," says developer Jonathan Durst, co-president of the Durst Organization. "But, now we find that people are willing to pay a premium [for sustainability]."

Durst is hesitant to say exactly how much more the tenants in the Helena's 497 market-rate units pay for their innovative environments. (In exchange for reserving 20 percent of the units for low-income tenants, the developer received tax-exempt financing.) Despite this reticence, Durst is well acquainted with trends in sustainable

THE HELENA'S GREEN STRATEGIES HELP SATISFY A GROWING DEMAND IN NEW YORK CITY'S REAL ESTATE MARKET.

real estate. His firm was one of the early adopters of large-scale green development. With the Helena's architect, Fox & Fowle, (now known as FXFOWLE), the Durst Organization designed and built Four Times Square, completed in New York City in 1999 and widely regarded as the country's first high-performance office tower.

Like the earlier Durst project, the only exterior signs of the aggressive environmental attributes are the building's

KEY PARAMETERS

LOCATION: New York, New York (Manhattan Island, Hudson River watershed)

GROSS SQUARE FOOTAGE: 600,000 ft² (55,700 m²)

COST: $100 million

COMPLETED: 2005

ANNUAL PURCHASED ENERGY USE (BASED ON SIMULATION): 55 kBtu/ft² (622 MJ/m²)

ANNUAL CARBON FOOTPRINT: (predicted): 15 lbs. CO_2/ft² (74 kg CO_2/m²)

PROGRAM: 580 studio, one- and two-bedroom apartments

TEAM

OWNER: The Durst Organization

ARCHITECT: FXFOWLE Architects with Harman Jablin Architects (residences) and B Five Studio (lobby)

ENGINEERS: Flack + Kurtz (MEP); Severud Associates (structural)

OWNERS REPRESENTATIVE: Rose Associates

ARCHITECTURAL CONSULTANT: Robert Fox, AIA

ACOUSTICAL: JRH Acoustical Consulting

ENVIRONMENTAL CONSULTANT: Allee King Rosen & Fleming

GREEN BUILDING CONSULTANT: e4 inc.

GENERAL CONTRACTOR/CONSTRUCTION MANAGER: Kreisler Borg Florman

« The Helena is clad with a high-performance window wall that includes trickle vents, allowing residents to let in a controlled amount of outside air without opening apartment windows.

The building's sun-filled spacious lobby is finished in terrazzo, mosaic tile, and mahogany.

photovoltaic (PV) panels. Along with an array that covers the southern and western faces of the penthouse mechanical enclosure, PVs integrated into the entrance canopy visible to passersby provide a modest 13.1 kilowatts (kW) of electricity, or 4 percent of the Helena's power requirements. The building's remaining electricity needs are met with utility-supplied electricity, but half of that is offset with a wind power purchase agreement.

The building relies on a combination of other technologies to achieve its energy savings such as individual water-source heat pumps in each apartment. One advantage of this system is that heat rejected from one apartment can be used to warm another, explains Charles Kryksman, a vice president at Flack + Kurtz, the project's mechanical, electrical, and plumbing engineers.

Water-source heat pumps are uncommon in New York City residential buildings where individual packaged terminal air conditioners, called PTAC units, are typically used. Although much less efficient than the Helena's water-cooled system, the through-wall PTACs are popular with landlords because of their low first cost and the ease of transferring the expense of their operation directly to tenants, says Kryksman.

Among the other energy-efficiency strategies are occupancy sensors that control lighting in corridors and stairwells. In each apartment a setback switch at the entrance allows residents to shut off all appliances plugged into the bottom receptacle of every outlet and all hard-wired lighting. Activating the switch also sets the apartment thermostat to 60 degrees in the winter and 85 degrees in the summer.

The Helena's most space-intensive green feature is the blackwater recycling system, which reclaims about 43,000 gallons of wastewater each day. The processed blackwater, along with storm water, is used in the cooling tower, for flushing toilets, and for irrigation of the 12,000 square feet of green roofs. The system occupies about 5,000 square feet on the north side of the second and third floors, an area not suitable for apartments because of the proximity of an existing structure. Although the equipment added about 1.5 percent to the construction cost of the building, the square footage did not count

The Helena's blackwater recycling system reclaims about 43,000 gallons of wastewater each day.

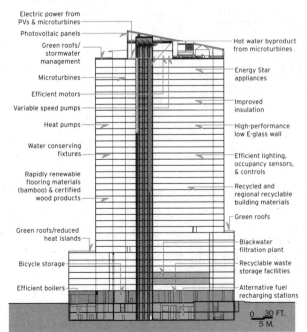

against the tower's allowable floor area ratio and made perfect use of space that would not generate revenue, points out Bruce Fowle, FAIA, FXFOWLE senior principal. The system is sized to also handle the blackwater load of the Rose, an apartment building the developer plans for the site just to the west of the Helena.

Urea formaldehyde-free wheatboard millwork, and paints, carpeting, and other finishes with no or low levels of VOC contribute to excellent air quality. Trickle vents, approximately six-inch-long operable slots in the aluminum frames of the high-performance window walls, give residents the option of letting a controlled amount of outside air into their apartments without opening windows—a good ventilation option especially in cold weather.

To help residents maintain the air quality inside their apartments, the Helena's management encourages residents to adopt green housekeeping practices. A welcome package distributed to new tenants includes information about the cleaning methods employed throughout the building's common areas and a supply of the same environmentally benign products used by the staff. It also contains instructions for operation of the appliances, information about the setback switch, and the schedule of the compressed natural gas fueled shuttle that runs between the Helena and a major subway stop a half-mile away.

Despite these measures, compliance with a Leadership

SECTION A-A

Electric power from PVs & microturbines
Photovoltaic panels
Green roofs/ stormwater management
Microturbines
Efficient motors
Variable speed pumps
Heat pumps
Water conserving fixtures
Rapidly renewable flooring materials (bamboo) & certified wood products
Green roofs/reduced heat islands
Bicycle storage
Efficient boilers

Hot water byproduct from microturbines
Energy Star appliances
Improved insulation
High-performance low E-glass wall
Efficient lighting, occupancy sensors, & controls
Recycled and regional recyclable building materials
Green roofs
Blackwater filtration plant
Recyclable waste storage facilities
Alternative fuel recharging stations

0 30 FT.
5 M.

Because the Helena is located on a former industrial site in a changing neighborhood on the western edge of Manhattan (above, left), many apartments have Hudson River views (above).

SKY CONDITIONS

In New York City's often cloudy climate, access to daylight is particularly welcome.

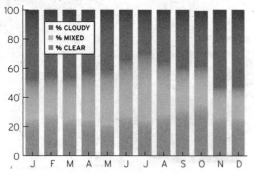

Legend:
- % CLOUDY
- % MIXED
- % CLEAR

TEMPERATURES & DEW POINTS

Temperatures range widely in New York City, with only limited swing seasons with temperate conditions.

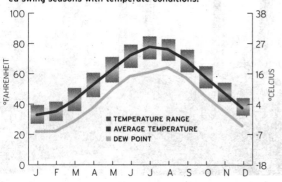

Legend:
- TEMPERATURE RANGE
- AVERAGE TEMPERATURE
- DEW POINT

HEATING/COOLING DEGREE DAYS

Both heating and cooling loads are significant, so it pays to have an HVAC system that can do both efficiently.

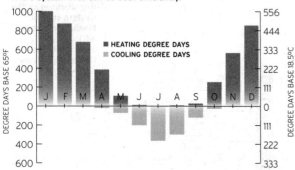

Legend:
- HEATING DEGREE DAYS
- COOLING DEGREE DAYS

A LEED prerequisite regarding tobacco smoke nearly derailed the project's certification.

TYPICAL FLOOR

1 Apartments
2 Balcony
3 Laundry

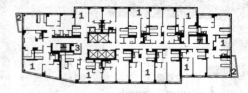

GROUND FLOOR

1 Retail
2 Lobby
3 Plaza entry
4 Canopy
5 Service
6 Trash
7 Bike storage
8 Parking ramp
9 Drive through

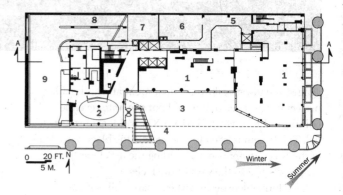

0   20 FT.
5 M.

Winter   Summer

in Energy and Environmental Design (LEED) indoor environmental quality prerequisite regarding tobacco smoke nearly derailed the project's certification by the U.S. Green Building Council (USGBC). The rating system requires that certified buildings be smoke free or that smoking rooms be negatively pressurized when compared with surrounding areas. Buildings that permit smoking must also comply with a testing standard created specifically for laboratories—a high bar for a residential building. "A lone smoking room in an office is easy to arrange," says Pamela Lippe, president of e4, the project's green building consultant. "But in a residential building, you can't tell people not to smoke, and it is [impossible] to make every apartment negatively pressurized relative to the adjacent one," she says.

Lippe recommended an alternative compliance path that included "supersealing" apartments with extra tap-

« Green roofs (far left) help with stormwater management. Photovoltaics integrated into the mechanical enclosure (middle) and the entry canopy provide about 4 percent of the building's power.

SITE DIAGRAM

1 The bulk of the building is pulled back from the street wall, preserving views to the Hudson River.

2 The streetwall is maintained at the building base.

3 Back of house functions are sited on the side of the building with less appeal.

4 Retail at the building base adds vitality to the street life.

5 A second phase of construction, an apartment building called "The Rose" is planned for an adjacent site.

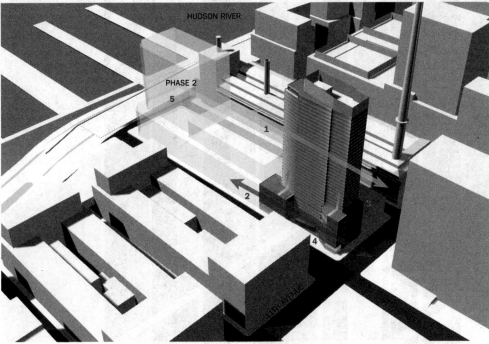

HUDSON RIVER

PHASE 2

SOURCES

BLAST-FURNACE SLAG IN CONCRETE MIX: Gran-Cem Cement

METAL AND GLASS WINDOW WALL: EFCO Corporation Model 890-I with AFGD Glass

GREEN ROOF: American Hydrotech 6125-FR and 6125-EV

PAINTS: Benjamin Moore Pristine Eco Spec and Moorecraft Super Spec

CARPET: Bentley Prince Street

ELEVATORS: Fujitec

MICROTURBINES: Ingersoll Rand

PHOTOVOLTAIC PANELS: altPower/GE Energy

ing and spackling, and gasketing and staggering outlets in walls between apartments. A blower door test is also required to demonstrate minimum air leakage from apartments. "This is just good construction but it rarely happens in practice," Lippe says.

The Helena did receive its LEED certification, achieving a Gold rating. However, even after certification, the owner and its consultants were still dealing with a few loose ends. Designers planned to use the waste heat generated by the building's two 70 kW microturbines to heat the domestic hot water supply. But activation of the turbines was long on hold, awaiting fire department final approval, along with the same units specified for other New York residential buildings. In mid 2007 the turbines were approved for use and have been in operation since, but Durst calls the delay the project's "biggest surprise." This relatively

minor glitch in the context of the otherwise successful project never affected the building's smooth operation, and since opening it has been a highly desirable place to live. The Helena has no vacancies and maintains a long waiting list for its apartments, says Durst.

Given the relative strength of Manhattan's residential rental market, it isn't surprising that Durst's company is betting that the rental boom will continue. Some 20 blocks south of the Helena, the developer completed another FXFOWLE-designed mixed-use tower in late 2007. The 458-rental-unit building is expected to achieve at least a LEED Silver rating. But even if the market stalls, Durst's business strategy should prove sustainable in every sense of the word. As Lippe points out, "to the extent the market begins to falter, [green] developers will be better positioned to keep their buildings full." «

CASE STUDY
PLAZA APARTMENTS
SAN FRANCISCO

JOANN GONCHAR, AIA

LEED SCORES
LEED-NC Version 2 Silver

SITES [SS]	6	14
WATER [WE]	2	5
ENERGY [EA]	8	17
MATERIALS [MR]	6	13
INDOORS [EQ]	8	15
INNOVATION [ID]	5	5

● POINTS ACHIEVED   ● POSSIBLE POINTS

A Room of One's Own

IN THE HEART OF SAN FRANCISCO'S SKID ROW, ONCE-HOMELESS ADULTS FIND SHELTER IN A SHOWPLACE OF GREEN DESIGN

A NEW RESIDENTIAL BUILDING IN A still-rough section of San Francisco's South of Market neighborhood is helping satisfy a set of ambitious social and environmental goals. The $22 million Plaza Apartments, open since early 2006, is one of the city's first green affordable housing projects. It incorporates sustainable materials, on-site power generation, and strategies to ensure good indoor air quality. It is designed to achieve a 25 percent energy savings over a building that complies with California's already-tough energy standard, Title 24.

First conceived as an apartment building for very low-income San Franciscans, the project was reconfigured in a late stage of construction—after eight of its nine-story exposed-concrete structure had been poured and much of the mechanical systems installed—to provide permanent housing for the chronically homeless. After the shift in program, designers maintained the

layout of the 106 studio apartments but were required to provide spaces for additional staff, such as social workers and health-care professionals. "Our goal is to provide everything residents need for fully independent living," says Erin Carson, the former project manager for the building's owner, the Public Initiatives Development Corporation (PIDC). The service staff grew from two to 10 people, according to Carson, now a development specialist with the parent organization of PIDC, the San Francisco Redevelopment Agency.

Despite the many support services offered on-site, the Plaza has a decidedly non-institutional flavor. The building's upper stories are clad in composite panels in a variety of earthy hues. The street level includes retail space and a lobby for a below-grade black-box theater to be leased by a Filipino performing arts company, which occupied the decrepit 1920s-era building formerly on the site at Sixth and Howard streets. "One of the goals of the project is to improve the neighborhood without displacing tenants," says Richard Stacy, AIA, principal of locally

« On the Plaza's facade, earth-hued rainscreen panels of recycled craft paper, wood veneer, and resin are contained within the exposed structural grid.

KEY PARAMETERS

LOCATION: San Francisco (Vista Grande watershed through the San Francisco Canal)

GROSS SQUARE FOOTAGE: 56,800 ft² (5,277 m²)

COST: $22 million

COMPLETED: January 2006

ANNUAL PURCHASED ENERGY USE (BASED ON SIMULATION): 58 kBtu/ft² (660 MJ/m²)

ANNUAL CARBON FOOTPRINT (PREDICTED): 11 lbs. CO_2/ft² (54 kg CO_2/m²)

PROGRAM: Studio apartments, mental and physical health-care facilities

TEAM

OWNER: Public Initiatives Development Corporation

ARCHITECT/INTERIOR DESIGNER: Leddy Maytum Stacy Architects and Paulett Taggart Architects

ENGINEERS: OLMM Consulting Engineers (structural); Telamon Engineering Consultants (civil); CB Engineers (mechanical); POLA Design & Engineering Services (electrical)

COMMISSIONING AGENT: Timmons Engineering

BUILDING ENVELOPE CONSULTANT: Simpson Gumpertz & Heger

GENERAL CONTRACTOR: Nibbi Brothers

TIM GRIFFITH

RESIDENTIAL **129**

The reception area
(above) looks out
onto a common
courtyard.

Windows on the
southwest side of
the building are
recessed to reduce
heat gain
(opposite).

based Leddy Maytum Stacy Architects. His firm, in association with Paulett Taggart Architects, also of San Francisco, designed the building.

Residents enter the Plaza Apartments through a leafy courtyard leading to a double-story entry hall flanked by administrative offices and a community room. The warm tones of bamboo floors, sustainably harvested cherry veneers, and ochre-colored walls contrast with exposed concrete and slate.

A typical residential floor contains 14 apartments, each a very compact 280 square feet, including a kitchenette and bath. The units are arranged in pinwheel fashion around the building's central core. The organization allows for windows and ventilation louvers at the end of each common corridor, providing daylight, views, and fresh air—amenities

that are particularly valued given the building's density, points out Carson.

Apartment finishes are simple and durable, including formaldehyde-free wheatboard casework, linoleum kitchen floors, and non-VOC paints. On the assumption that the carpeting would need to be entirely replaced whenever a resident moved, the architects chose a rolled material with a high percentage of recycled content. For the corridors, they selected carpet tiles so that maintenance staff could easily replace stained or worn portions.

On the exterior, the corridor windows create a vertical slice through each facade, interrupting the exposed structural grid established by the width of each apartment. Within this grid, high-performance operable window walls alternate with the composite panels made of recycled craft paper, wood veneer, and resin.

The pinwheel arrangement of the apartment floors allows for windows and ventilation louvers at the end of corridors, providing daylight, views, and fresh air.

These panels, part of an open-jointed rainscreen facade system, provide a number of performance advantages, according to Stacy. For example, the wall assembly includes batt insulation between studs as well as a layer of rigid insulation on the outboard side of the framing, enhancing the thermal properties of the building envelope. In addition, the approximately 1-inch cavity between the panels and the building's weather-resistant barrier allows air to circulate behind the facade, mitigating solar gain, he says.

Concern about solar gain was one of the reasons the architects gave slightly different expression to each facade. On the southwest side of the building, the panels are flush with the structure, but the windows are recessed, providing shading from the hot afternoon sun. However, on the southeast facade, where thermal gain is less of a worry, the relationship is reversed. In addition to tuning the facades to their respective orientations, the configuration creates subtle variations in the play of light, shadow, and materials.

On the roof is a 28 kW photovoltaic (PV) array that generates about 5 percent of the Plaza's power needs. To provide this amount of electricity, the building's compact roofscape is almost completely covered with PV panels,

LIVING UNIT AXO

1 Living space carpeting with 45 percent post-consumer recycled content
2 Linoleum kitchen floor
3 Rubber bathroom floor
4 Non-VOC paint
5 Hydronic heating system
6 Framed wall with Z-shaped duct
7 Rigid insulation
8 Rainscreen cladding system
9 Low-E glazing system with operable window
10 Formaldehyde-free wheatboard casework

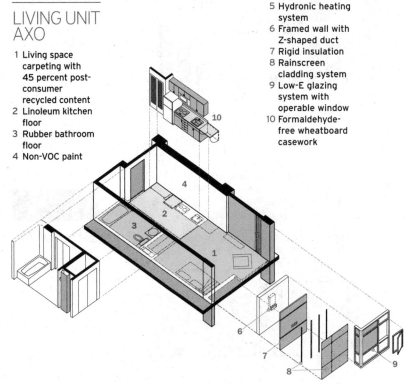

SKY CONDITIONS

Skies are relatively clear year-round after the fog burns off, creating a direct solar load that must be carefully managed.

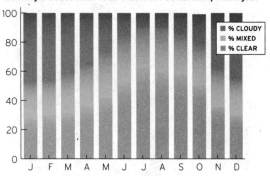

Legend:
- % CLOUDY
- % MIXED
- % CLEAR

TEMPERATURES & DEW POINTS

Daily and seasonal temperature variations are quite small.

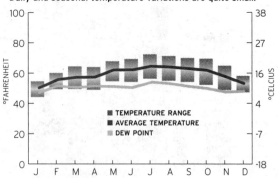

Legend:
- TEMPERATURE RANGE
- AVERAGE TEMPERATURE
- DEW POINT

HEATING/COOLING DEGREE DAYS

Moderate temperatures provide comfortable outdoor conditions much of the year.

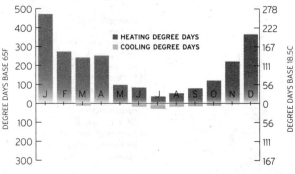

Legend:
- HEATING DEGREE DAYS
- COOLING DEGREE DAYS

NATOMA STREET
SIXTH STREET
HOWARD STREET

Summer / Winter / N

SITE PLAN

Visitors and residents enter the building from Howard Street through a landscaped and paved courtyard. All of the Plaza's ground-floor common areas face this outdoor gathering space.

creating a logistical challenge for the mechanical engineer, who needed to provide clearance for the many plumbing vents and bathroom fans and ensure access to mechanical equipment. "The roof is just jam-packed," says Chikezie Nzewi, project engineer for CB Engineers, San Francisco, the building's mechanical consultant.

Given San Francisco's temperate climate, mechanical cooling was deemed necessary only in the retail area and the theater. Those spaces are served by water-cooled heat pumps that receive condenser water from a roof-mounted cooling tower. A radiant hot-water system provides heating for most of the remainder of the building. Two natural-gas-fired boilers with thermal efficiencies of 85 percent generate the hot water, and pumps with variable-speed drives circulate it throughout the building.

Nzewi credits close coordination among team members for achieving the best layout for the hydronic system's infrastructure. He worked with the designers to precisely locate each radiator, saving an enormous amount of piping and making a comfortable furniture arrangement possible in the tight apartments. "Usually architects just provide an open shaft," he says.

Because California code limits permissible noise transmission levels from outside to inside, the designers provided another apartment ventilation source in addition to the operable windows. Through-wall Z-shaped transfer grilles let in outdoor air but prevent the transmission of traffic noise and other street sounds into the residences, while the bathroom "scavenger" fans provide constant low-level air changes. "From an energy standpoint, the arrangement is a bit of a negative, but it does provide for good indoor air quality," says Stacy.

In November 2007, the project was awarded Silver certification through the U.S. Green Building Council's Leadership in Energy and Environmental Design (LEED) rating system. However the path to

« Finishes in the compact apartments (far left) are simple and durable, including formaldehyde-free wheatboard casework, linoleum kitchen floors, and carpeting with a high percentage of recycled content in the living area. A common room opens off a double-story entry lobby (left). In these spaces, the warm tones of bamboo floors, sustainably harvested cherry veneers, and ochre-colored walls contrast with exposed concrete and slate.

NINTH FLOOR

1 Apartments
2 Laundry
3 Deck
4 Lounge

GROUND FLOOR

1 Courtyard
2 Lobby
3 Reception
4 Office
5 Flex space
6 Entry hall
7 Community room
8 Kitchen
9 Pantry
10 Trash room
11 Theater lobby
12 Retail

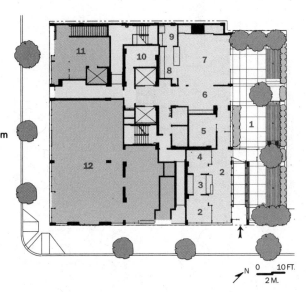

N   0   10 FT.
‾‾‾‾‾‾‾‾‾
2 M.

certification was not smooth. As part of an alternate compliance path to a LEED prerequisite regarding tobacco smoke, the owner commissioned a blower door test. The procedure is used to demonstrate minimum air leakage from one apartment to another and into common corridors. But the test was performed incorrectly giving erroneous results that indicated significant leakage, putting certification in jeopardy. The project passed a subsequent test without difficulty.

Though certification was somewhat arduous, team members say they found the process valuable. The need to demonstrate compliance helped ensure that performance goals were met. "A lot of things just won't get done if you don't go through certification," says Paulett Taggart, FAIA. She credits the documentation process with helping the team even exceed some of the goals established at the project's outset. For example, the contractors recycled almost 90 percent of the demolition debris even though the official requirement was only 75 percent. "I don't think we would have achieved that if there wasn't someone keeping a close watch," she says. «

SOURCES

METAL/GLASS CURTAINWALL: Kawneer

WOOD EXTERIOR CLADDING AND RAINSCREEN PANELS: Parklex

WINDOWS: PPG, Solexia, Industrex, Superlite II

DOORS: Marshfield Signature Series, FSC Certified (wood); Forderer Cornice Works (metal)

ROOFING: Johns Manville

PAINTS AND STAINS: ICI Dulux Lifemaster 2000 Interior, No VOC paint; Minwax (wood stain), L&M Construction Chemicals Dress & Seal WB, Concrete Graffiti Control, Prosoco Graffiti Barriers Block-Guard, low VOC

LIGHTING: Indessa, Prudential, Delray, Lightolier, Vibia, BK Lighting, Kim

CARPET: Tandus Collins & Aikman, Patcraft EcoSol

FURNISHINGS: Steelcase, Ecodura, Brayton, Howe

PHOTOVOLTAIC PANELS: Mitsubishi

RESIDENTIAL BATHROOM FLOORS: Nora Rubber

CASEWORK WHEATBOARD: Dow Woodstalk

Chapter 7

SCIENCE/TECHNOLOGY

APPLYING SUSTAINABLE DESIGN PRINCIPLES TO SCIENCE AND RESEARCH institutions presents a bit of a conundrum. On the one hand, the people who work in these buildings tend to be inquisitive about the world, demanding of high-quality environments in which to work, but also concerned with how those facilities use energy and natural resources. On the other, laboratory and health-care buildings tend toward the extreme end of energy use in buildings, with many code-mandated, energy-intensive requirements for air changes per hour and varying air zones of negative and positive pressures. Balancing those needs falls to architects, engineers, and contractors, who must work together to achieve an integration of building components that not only meets these demands, but also stays within tightly fixed institutional budgets.

What is remarkable about the four case studies that follow is their success in integrating what are arguably some quite novel sustainable design strategies—two of these projects achieved LEED Platinum—that don't compromise the program ambitions of the clients. The Oregon Health and Science University's Center for Health and Healing in Portland incorporates a wide range of uses—fitness centers, laboratories, exam rooms, a surgery suite—but achieved a Platinum rating through using a segregated mechanical system that shaves 60 percent off the state's energy code. Imagine if that were a baseline for all medical buildings! The Global Ecology Center in Stanford, California, saves energy by dividing the air-supply system from the water-based system for heating and cooling. And somewhat radically, the designers installed a rooftop water-spray system that uses colder nighttime air to chill water, collecting it to cool the building during the next day. The U.S. Department of Energy's Molecular Foundry in Berkeley, California, reduced by 28 percent the state's Title 24 energy requirements by "right-sizing" the mechanical system and installing variable-frequency drives for air-handling equipment and laboratory fume hoods. The Hawaii Gateway Energy Center at NELHA, Kailua-Kona, Hawaii—another LEED Platinum project—has a mechanical system that relies on pumping cold seawater into coils that can then be used to condense water from Hawaii's humid air. Like the scientific processes practiced daily in many of these facilities, the sustainable designs employed in these projects represent a new understanding of how our buildings must engage with the earth's natural forces. «

LEED SCORES
LEED-NC Version 2 Platinum

SITES [SS]	13	14
WATER [WE]	5	5
ENERGY [EA]	14	17
MATERIALS [MR]	8	13
INDOORS [EQ]	10	15
INNOVATION [ID]	5	5

● POINTS ACHIEVED ● POSSIBLE POINTS

CASE STUDY

CENTER FOR HEALTH AND HEALING
PORTLAND, OREGON

RANDY GRAGG

A Healthy Dose of Green

A UNIVERSITY'S MEDICAL OFFICE BUILDING ACHIEVES LEED PLATINUM THROUGH SOME SERIOUS INTEGRATED DESIGN STRATEGIES

At the first meeting with the architects designing the Oregon Health and Science University's Center for Health and Healing, developer Dennis Wilde posed a challenge: Reduce the capital costs for the building's mechanical systems by 25 percent but make it outperform the Oregon energy code by 60 percent.

For a simple, single-use building, Wilde's goal would have been bold enough. But given the center's unprecedented mix of swimming pools, a surgery suite, exam rooms, offices, and research labs—each with heating, cooling, and ventilation demands far beyond the norm—it was nothing short of audacious. "It was impetuousness, plain and simple," recalls Wilde, a principal at Gerding Edlen. "We habitually build buildings full of mechanical equipment that's seldom used. Why the hell not get creative?" What the team quickly discovered is that the

proverbial sum, in fact, could be much greater than the parts, particularly if you make sure most of those parts serve more than one function.

Designed by GBD Architects in close collaboration with Interface Engineering, the Center for Health and Healing is a lesson in the architecture of integration. "The more we started really looking at the systems, the more we were able to cut costs," says GBD's lead designer, Kyle Andersen, AIA. "It was about reducing the equipment or rethinking it to do multiple things, instead of just looking through the tables and picking what's always worked in the past."

Yet, the design of the $145-million, 400,000-square-foot center needed to do far more than save energy as the first building in the university's new 10-acre satellite campus and one of the first in Portland's largest urban redevelopment in 40 years: the 38-acre, $1.9-billion River Blocks development. It would rise 16 stories next to a new streetcar line to downtown and a stylish new aerial tram linking the district to the university's main hilltop campus, a 3,000-foot flight away. Housing the district's

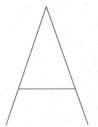

KEY PARAMETERS

LOCATION: Portland, Oregon (Willamette River watershed)

GROSS SQUARE FOOTAGE: 400,000 ft^2 (37,000 m^2) finished; 262,000 ft^2 (24,300 m^2) underground parking

COST: $120 million (construction only), $160 million (including FFE's)

COMPLETED: October 2006

ANNUAL PURCHASED ENERGY USE (BASED ON SIMULATION): 102 kBtu/ft^2 (1,160 MJ/m^2), 61 % reduction from base case

ANNUAL CARBON FOOTPRINT (PREDICTED): 22 lbs. CO$_2$/ft^2 (109 kg CO$_2$/m^2)

PROGRAM: Entry atrium, cafe, pharmacy, retail eye clinic, day spa, wellness center, conference center, imaging, ambulatory surgery, outpatient clinics and offices, educational offices, and research laboratories

TEAM

OWNER: RIMCO: OHSU Medical Group

DEVELOPER: Gerding Edlen Development

ARCHITECT/INTERIOR DESIGNER: GBD Architects

CONSULTING ARCHITECT/MEDICAL CONSULTANT: Peterson Kolberg & Associates

LANDSCAPE: Walker Macy

ENGINEERS: KPFF Consulting Engineers (structural); Interface Engineering (MEP/commissioning); OTAK (civil); GeoDesign (geotechnical)

ENVIRONMENTAL CONSULTANT: Brightworks Northwest

LABORATORY PLANNING: The Estimé Group

ACOUSTICAL: Altermatt Associates

POOL CONSULTANT: Aquatic Design Group

ENVELOPE CONSULTANT: The Facade Group

GENERAL CONTRACTOR: Hoffman Construction

JAMZ PHOTOGRAPHY

Radiant floors in the lobby offset heat gain from the large expanses of glass (above). The first floor lobby includes low-VOC finishes (top right). The second floor's fitness center includes the lap pool (above right).

first health club and all of the university hospital's outpatient services, the center would be a gateway building that officials wanted to stand as an "icon of health."

But the steepest challenge, according to Andersen, was the complex stack of uses inside the building: wellness, fitness, and physical therapy facilities, plus a conference center on the lower floors; outpatient clinics, imaging, and ambulatory surgery on the middle floors; and offices and laboratories on top. With no major foundation or government grants or private benefactor, the university doctors' group was developing the center, in effect, as a "spec med science facility" with Gerding/Edlen as the turnkey developer. As pro forma driven as any spec office building, every use inside had to pay its own way based on either future expected fees or rents.

The resulting tower stands with pragmatic simplicity: a stack of programs rising between stair towers with a three-story glass-box atrium lobby fronting the busy aerial tram station. With the center's near-perfect compass

orientation, computational fluid dynamic modeling showed the building could be ventilated almost entirely through passive means. The north side features a ventilation system that draws air through the building, its circulation given a boost by the heat of lights and computers. The stair towers at the building's east and west ends both reduce the building's solar loads while functioning as stacks to further draw air out.

The team's constant search for "double duties" in the design is most clearly visible in the south elevation— what Anderson calls the "machine side"—which rises in a unitized curtain wall fitted with sunshades, each equipped with photovoltaic panels that add up to 60 kW of power. The sunshades alone reduced the building's cooling loads by 30 tons, enough, according to Andersen, to pay for the brackets used to mount them.

Daylighting studies the team conducted to determine the sunshades' size and placement revealed another opportunity: the south side's sun-baked upper floors

With 55 LEED points, it's the largest health-care facility in the country so far to earn a Platinum rating.

were perfect for a solar collector. Hence, Andersen stepped the top of the building back five feet for a greenhouse-like space clad in low-iron glass behind which the 100-degree-plus sunny-day temperatures preheat water circulated to the labs, swimming pools, and the lobby and waiting area's radiant floors. The collector is projected to provide about one percent of the building's overall energy, which, according to Andersen, will lead to a capital cost payback period of only nine years. "This building spends a lot of energy heating up water," Andersen notes. "We looked for any way we could do it cheaply or with a quick payback." Even the five natural-gas-fired microturbines, designed to produce 30 percent of the building's energy, provide excess heat used to warm the swimming pool.

The team's efforts to curb water usage and disposal grew as ambitious as Wilde's energy goals. Green roofs and an on-site bioreactor were designed to process every drop of rainwater, groundwater seepage, and sewage on site, to then be reused for landscape irrigation, toilet flushing, or radiant cooling, as well as to charge a bioswale that seeps into the nearby Willamette River. Overall, the building's resource-use modeling proposed a possible 68 percent savings in water usage, or about 2.1 million gallons annually. For all their con-

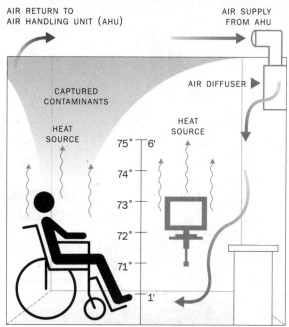

The center connects to an aerial tram (at right in the above picture) that ties into a larger transportation system serving Portland.

AIRFLOW DIAGRAM

The displacement ventilation scheme for examination rooms ensures that contaminated air is exhausted from treatment areas and therefore won't infiltrate adjacent spaces.

SKY CONDITIONS

In spite of the inefficiencies due to cloudy winters, solar-thermal collectors proved viable because hot water demand is high in summer as well.

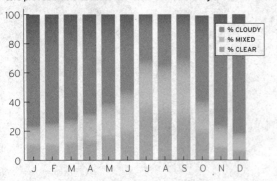

- % CLOUDY
- % MIXED
- % CLEAR

TEMPERATURES & DEW POINTS

Relative mild temperatures year-round make natural ventilation and passive cooling attractive.

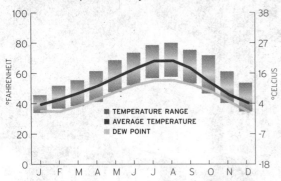

- TEMPERATURE RANGE
- AVERAGE TEMPERATURE
- DEW POINT

HEATING/COOLING DEGREE DAYS

In a large, equipment-laden building like OHSU, cooling is the predominant load even though the climate shows more heating degree days.

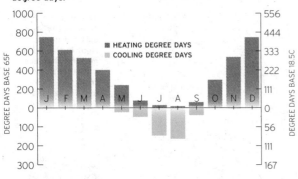

- HEATING DEGREE DAYS
- COOLING DEGREE DAYS

SITE PLAN

1 OHSU Center
2 Interstate 5
3 Willamette River

WATER SYSTEM DIAGRAM

The center has four separate water systems, including a blackwater system that feeds a non-potable water supply, a conventional potable water system, and rainwater collection system that feeds the fire water cistern, as well as the mechanical system.

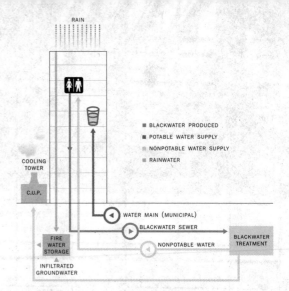

- BLACKWATER PRODUCED
- POTABLE WATER SUPPLY
- NONPOTABLE WATER SUPPLY
- RAINWATER

centration on integrating the building's systems, Andersen and his team designed an unusually comfortable and inspired space. The center's most dramatic feature is the atrium. Here, energy and cost savings served aesthetics since the underground parking garage's ventilation system also exhausts the atrium in the event of a fire, leaving the ceiling free of bulky fans. With all the clinical and surgery waiting rooms located on the building's north side—and all air-conditioned with radiant floors and chilled beams—the center's main lobby and waiting areas are unimaginably quiet and calm for a major medical clinic. Gracious north windows flash views of the fitness equipment and basketball courts to passing aerial tram and streetcar riders.

Sadly, the building lost what might have been its most iconic features: cowels to more powerfully vent the stair towers and wind turbines to provide electricity. The building's doctor/investors were willing to shoulder the 30-year payback period, Andersen recalls, but the Portland Design Commission voted down the necessary height variance.

So how well has the center measured up to Wilde's challenge? Wilde says the building's systems were "cost-neutral with no savings, but no huge premium." Interface's lead engineer, Andy Frichtl, PE, argues the goals were met, if you calculate the internal rates of return on capital paybacks due to energy saved over the building's life. The energy modeling shows the center should operate at 61 percent below energy code. One thing is for sure: with 55 LEED points, the center became the largest health-care facility in the country so far to earn a Platinum rating.

But commissioning the center has proven complex, taking over 10 months, according to chief building engineer Mark Schnackenberg. The optimistic models, he notes, were based on more typical, 9-to-5 building usage, not on laboratories in which a researcher might want to work all night with the building's systems pumping 16 hourly changes of air into a lab. The newly-planted green roofs and landscapes have demanded more water than anticipated, Schnackenberg says, leaving the building's cistern too empty at times to flush the toilets. That triggered the back-up system along with the 17-cents-per-gallon penalty the city demanded for the accelerated per-

14TH FLOOR

1 Lab
2 Typical office
3 Conference room
4 Small labs
5 Open offices

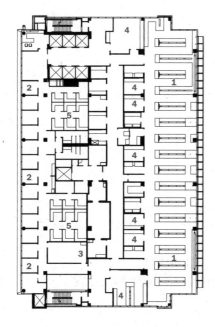

2ND FLOOR

6 Gymnasium
7 Physical therapy
8 Lap pool
9 Spa pool
10 Therapy pool
11 Kitchen
12 Locker rooms

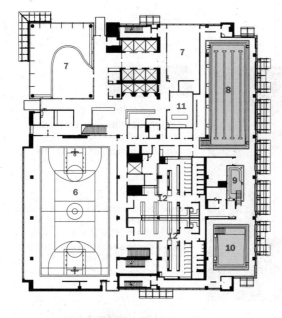

1ST FLOOR

13 Lobby
14 Cafe
15 Retail
16 Coffee shop
17 Spa
18 Cardio fitness
19 Fitness room
20 Loading dock
21 Pharmacy

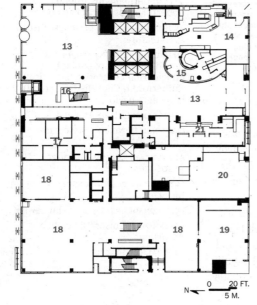

N ← 0 20 FT.
5 M.

« Patient exam rooms include daylighting and access to views, as well as a displacement ventilation scheme that keeps indoor air quality levels high.

mitting of a previously untried system—about $250,000/year—according to Schnackenberg. The bioreactor had to be upsized to handle the surprising large clinical and laboratory loads and has frequently broken down due to rags and other debris tossed in by busy technicians failing to follow disposal rules. "In retrospect," Schnackenberg says, "a bioreactor probably wasn't the right fit for this kind of facility."

Engineer Frichtl relishes the "software-like" complexity of the systems. Developer Wilde longs for the simplicity achieved with the solar collector—but throughout the building's other systems, too. The center, he contends, is overly complex, when the future design of more sustainable buildings should aim "to be smarter and simpler." But the center's operating engineer, Schnackenberg, wants to wait and see before judging. He says the center has yet to hit any of the projected efficiencies. "But it will take a full year of operating fully commissioned every season to really know," he adds. "This is uncharted territory. We're really just starting." «

SOURCES

GLASS: Viracon; VE1-2M; insulated-panel glazing: Kalwall

DOORS: Lynden (agri-core doors)

LOW-SLOPE ROOFING: American Hydrotech fluid applied membrane at eco roofs

PAINTS AND STAINS: PPG, Bona Sport

BAMBOO PANELING: Smith and Fong

FLOORING: Forbo Flooring–Marmoleum; Hardwood Flooring/Wood Gym Flooring: Robbins Hardwood Floors (from Armstrong); Hardwood Flooring: EcoTimber; Resilient Athletic Flooring: Dodge-Regupol–ECOsurfaces

TACK FABRIC: Maharam, Knoll

WOOD BENCHES: Freeman Corporation; Endura Wood Products

ELEVATORS: Otis Gen2

LIGHTING: Guth Lighting Enviroguard; Ledalite Pure FX; Mark Lighting Magellan; Kurt Versen Square; Lightolier Calculite

PHOTOVOLTAICS/SUNSHADES: Sharp (support structure by Benson Industries)

CARPET: Shaw Industries–Drops; Urban Grid; InterfaceFLOR Commercial–Lowes

BIOREACTOR: Mattsson Environmental Solutions

CHILLERS: York

CONTROLS: Alerton Building Management System; Wattstopper; Square D Powerlink; PCI Controlkeeper

GLOBAL ECOLOGY CENTER
STANFORD, CALIFORNIA

Planetary Perspectives

DESIGN FOR LABS AND OFFICES FOR A TEAM OF CLIMATE RESEARCHERS MIMICS NATURAL SYSTEMS TO DRIVE DOWN ENERGY USE AND CARBON EMISSIONS

NADAV MALIN

⩘ **Readily accessible conference rooms complement the open-plan offices by providing private space for meetings.**

HIS WAS THE FIRST program I've seen in which you can tell that someone approached the building with sustainability in mind," says Scott Shell, of EHDD Architecture, in reference to the client's concept document for the Department of Global Ecology, a new arm of the Washington, D.C. –based Carnegie Institution. Located alongside the venerable Department of Plant Biology on a 7.4-acre site leased from Stanford University, Global Ecology has 50 researchers and staff who study planetary systems, especially the changes, including those affecting climate and biodiversity. "We're concerned about humanity's effect on the planet," says director Chris Field, "particularly regarding energy use." That concern came through loud and clear in their priorities for the facility.

Rather than clearing a mature oak forest from the site to create a one-story structure, the designers chose to tuck the building into a previously paved utility area at the back of the property, creating a new core for the campus. A two-story building better accommodated the area's smaller size, and the narrow, 40-foot-wide plan facilitated daylighting.

The program called for roughly equivalent amounts of lab and office space. Instead of adopting a typical approach, giving each research team offices adjacent to its labs, designers put all the labs on the first floor and the offices above, a decision that enhanced both interaction and flexibility. This separation also saves energy because it lets large amounts of outside air into the lab zone without overventilating the offices.

KEY PARAMETERS

LOCATION: Stanford, California (San Francisco Bay watershed)

GROSS SQUARE FOOTAGE: 10,890 ft² (1,000 m²)

COST: $4 million

COMPLETED: March 2004

ANNUAL PURCHASED ENERGY USE (BASED ON SIMULATION): 111 kBtu/ft² (1,260 MJ/m²), 63% reduction from base case

ANNUAL CARBON FOOTPRINT: (predicted): 20 lbs. CO_2/ft² (97 kg CO_2/m²)

PROGRAM: Lab, office

TEAM

OWNER: Carnegie Institution Department of Global Ecology

ARCHITECT/INTERIOR DESIGNER: EHDD Architecture

ENGINEERS: Rumsey Engineers (mechanical and plumbing); Engineering Enterprise (electrical); Rutherford & Chekene (structural); BKF Engineers (civil)

LANDSCAPE: Lutsko Associates

LIGHTING: JS Nolan + Associates Lighting Design

ACOUSTICAL: Charles M. Salter Associates

LABORATORY DESIGN CONSULTANT: Design for Science

COST CONSULTING: Oppenheim Lewis

DAYLIGHTING: Loisos/Ubbelohde Associates

GENERAL CONTRACTOR: DPR Construction

Large bifold doors open to the courtyard and landscape on two sides, making the lobby a place for scientists to enjoy their immediate surroundings as they study the environment on a larger scale.

⌃⌃
A large clear-span
on the second floor
facilitates
daylighting and
natural ventilation,
and makes it easy
to reconfigure the
office space.

The department has many climate researchers on staff, so the designers felt it natural to develop energy systems related to their client's work. An evaporative downdraft chilling tower cools the lobby, working in a similar way to the katabatic winds that form as the temperature drops and moves air down the faces of glaciers. And in lieu of a large chiller, water is cooled by spraying it onto the roof at night, where it releases its heat through night-sky radiation; the cooled water is then stored in an insulated tank until it is needed. This system, originally developed by Richard Bourne of Davis Energy Group, has been so successful that the building uses a small backup chiller for additional cooling only rarely, and then only under peak conditions.

Radiant heating and cooling is delivered throughout the building via water pipes in the slab floors; air distribution is used exclusively for ventilation. Labs are typically ventilated by a mechanical system that can supply 100 percent outdoor air. When the air supply is also the delivery mechanism for heating and cooling, poor outside air temperatures can mean huge expenditures of energy to cool or heat air to comfortable levels. The department's water-based system saves energy by separating heating and cooling from ventilation requirements.

Placing several heavy-duty freezers for the labs in a semi-conditioned warehouse next door instead of in the labs themselves is another example of how the building program was developed with sustainability in mind. These deep-freezers produce a large amount of heat as they maintain temperatures as low as −80° Celsius, so keeping them out of the occupied building reduced the cooling load significantly.

COOLING TOWER

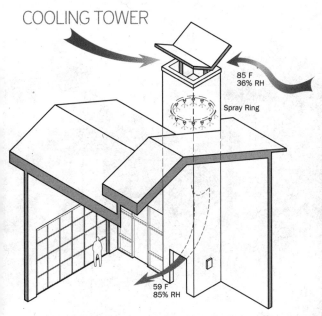

85 F
36% RH

Spray Ring

59 F
85% RH

⌃
A down-draft cooling tower provides direct evaporative cooling to the lobby when conditions permit.

⌃
Inside the cooling tower scientifically-sized droplets of water are released. As these evaporate in the warm, dry air, they cool the air, which then sinks to the bottom of the tower and escapes into the lobby. As the cool air exits, more warm air enters at the top.

Although the designers were sold on open offices as a way to optimize daylighting and natural ventilation, they questioned how to manage acoustics, a particular concern because much of the staff moved from private offices. Ultimately, Shell credits the occupants with making the building work. "They meet and frequently discuss its performance," Shell reports. On an occupant survey concluded by the Center for the Built Environment (CBE) of the University of California, Berkley, the facility received the second highest rating of the 158 buildings in the CBE database. And it was "the [only] green building that scored positively for acoustics," says Shell.

The natural ventilation also took some teamwork to figure out. "When they first moved in, they were trying to figure out when to open the windows," Shell says. Eventually they learned to keep the windows closed on really hot days, because the radiant cooling in the slab couldn't keep the space comfortable with the windows open.

Atmospheric impact was a serious consideration in material selection. The designers, looking to reduce the carbon emissions associated with certain materials, specified high levels of flyash in the concrete, which reduced the amount of cement used by more than half. Slabs for three adjacent greenhouses poured before the main building was constructed were used to test whether moist-curing was needed for the mix; results showed that applying a standard curing compound would be sufficient. In the main building, the high-volume flyash concrete posed a problem only when thin topping slabs were poured in cold weather. (These mixes don't produce as much heat as standard concrete, so it takes longer than usual before they are ready for finishing.)

Much of the building's equipment and materials,

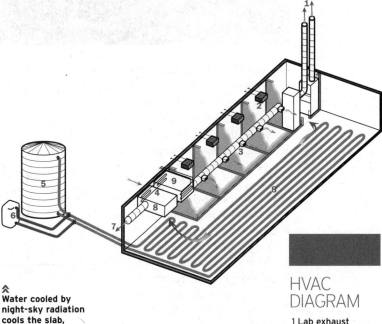

⌃
Water cooled by night-sky radiation cools the slab, while air-to-air heat exchangers reduce the energy needed to condition ventilation air for the laboratories.

HVAC DIAGRAM

1 Lab exhaust
2 Fan coils to night sky/boilers
3 Low-pressure ventilation duct
4 Flat plate heat exchanger
5 Night sky chilled water tank
6 Condensing boiler
7 Radiant heating and cooling
8 Exhaust fan
9 Air-handling unit supply fan

>> Sprinklers spray water onto the roof at night where night-sky radiation cools the water down enough for it to be used the next day to cool the building.

including siding, casework, workstation tabletops, sinks, and faucets, were salvaged, including both used and new items from off-spec orders found on the California Materials Exchange. "I was a skeptic before this project," admits Shell, about using salvaged materials, but "it was a lot easier than I expected. I'm now convinced that it is possible to do this even on larger projects."

In addition to its aggressive sustainability goals, Carnegie was interested in publicity for the project in hopes that it would inspire others to pursue similarly low-impact design. Nevertheless, it chose not to pursue LEED certification, for two reasons. The first was cost, which Field estimated to be in the tens of thousands of dollars. "We wanted to invest those funds in additional green features rather than in certification," he says. The second was that Carnegie had a specific set of environmental priorities for the project that didn't align precisely with those in LEED.

Unlike some owners, for whom a decision not to pursue LEED might have resulted in an unraveling of green strate-

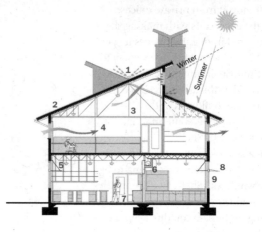

HEATING & COOLING DIAGRAM

1 Night spray radiant cooling
2 Spectrally selective roofing
3 Fully daylit interiors with lighting controls
4 Naturally ventilated top floor
5 Lightshelves
6 Efficient ventilation with heat recovery
7 Radiant slab heating and cooling
8 Sunshades
9 High-performance glazing

SKY CONDITIONS

The relative abundance of clear sky conditions, especially in the summer, enhances the potential for night-sky radiant cooling.

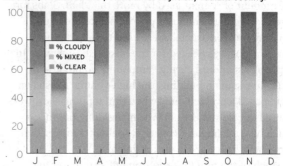

■ % CLOUDY
■ % MIXED
■ % CLEAR

gies as the project unfolded, this project's environmental performance remained a top priority. Without the LEED framework, however, there was no commissioning requirement, and the designers were unable to convince Carnegie to engage in a formal commissioning process. "I refused to accept the idea that the sign-off meant that they hadn't checked out this stuff," says Field. With or without commissioning, neither Carnegie nor the designers were satisfied until everything was working properly, which, for certain problematic systems, took a long time. "We did our commissioning on an item-by-item basis, rather than comprehensively," notes Field.

Although Field feels formal commissioning is unnecessary, the design team has no doubts. "I'm not going to do another green building unless it's commissioned, since the onus falls on the designers to prove the design wasn't at fault," says mechanical engineer Peter Rumsey. "When you do a green building and something goes wrong, people blame it on what's different and new. Our company paid for the commissioning ourselves, three times over," he concluded.

"I'm thrilled with how the technology works, now that we've worked out the kinks," says Field. Rumsey is also proud of the results, although he has learned that using a rooftop spray system for night-sky radiant cooling is easier on a flat roof, where the spray nozzles are readily accessible, than on a sloped roof. The main lesson, according to Rumsey: "It's possible to design a building that uses significantly less energy but is also very comfortable. People love being in there." ◀◀

SOURCES

WINDOWS: Kawneer 8225T-L

GLASS: Viracon 1-2M clear insulating glass with low-E coating

METAL ROOF COATING: BASF Ultra-Cool

CABINETWORK AND CUSTOM WOODWORK: FSC Certified Ash veneer; Columbia Forest "Europly" substrate

ZERO VOC PAINTS AND STAINS: Benjamin Moore, Frazee

FLOORING: Armstrong (linoleum)

CARPET: Interface (carpet tiles)

AMBIENT LIGHTING: Zumbotel Staff "Claris" with Lutron Hi-Lume Dimming ballast and Osram Sylvania T5HO lamps

CONTROLS: Lutron Continuous daylight dimming control system

PLUMBING AND FIXTURES: Falcon Waterfree urinals

Caroma Dual-Flush toilets

Electric hand dryer: Excel Dryer, Inc. XLerator Hand Dryer

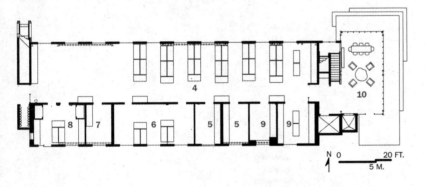

SECOND LEVEL

1 Private offices
2 Conference room
3 Open work area

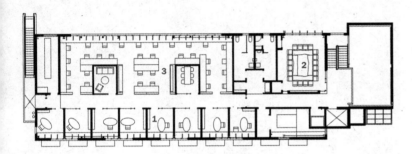

ENTRY LEVEL

4 Open work area
5 Clean work
6 Equipment room
7 Instrument room
8 Fume hood room
9 Procedure alcove
10 Lobby

N 0   20 FT.
        5 M.

TEMPERATURES & DEW POINTS

In spite of the comfortable climate, summer temperatures get high enough that cooling had to be taken seriously by the designers.

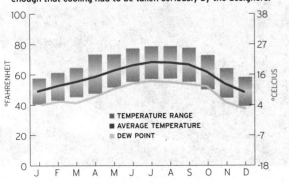

HEATING/COOLING DEGREE DAYS

Overall heating and cooling requirements in Stanford are lower than in most other U.S. locations.

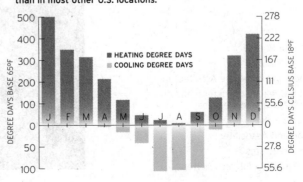

Doing Small Things in a Big Way

A NATIONAL CUTTING-EDGE LABORATORY FOR NANOTECHNOLOGY RESEARCH

A NEW, LEADING-EDGE NATIONAL research laboratory at Lawrence Berkeley National Laboratory (LBNL) took its inspiration from the foundries that ushered in the industrial revolution of the 19th century. Unlike these earlier foundries, however, the Molecular Foundry houses researchers who operate at a scale of nanometers (billionths of a meter), developing what some believe could be another revolution in technology and materials science.

Nanotechnology can involve such disparate fields as chemistry, physics, biology, computational science, materials science, and electrical engineering to develop everything from coatings and photovoltaic cells to ultra-fast computers and medical devices. LBNL's Molecular Foundry is one of five U.S. Department of Energy centers of nanoscale science research either completed or under construction and the only one on the West Coast.

Seeing the foundry, one is immediately struck both by the beauty of the location—on a steep hillside within the sprawling 200-acre LBNL research campus overlooking the city of Berkeley and San Francisco Bay beyond—and the success with which the designers integrated this dramatic building into the challenging site. The six-story building is on a 35-percent grade (dropping 70 feet vertically in 200 linear feet) and includes ground-level entrances on three floors. "The significant slope of the hillside site was a challenge," says project manager Suzanne Napier, AIA, of SmithGroup.

KEY PARAMETERS

LOCATION: Berkeley, California (Strawberry Creek and San Francisco Bay watersheds)

GROSS SQUARE FOOTAGE: 95,690 ft² (8,890 m²) including adjoining, two-story utility plant

COST: $52 million (total project cost, including equipment: $85 million)

COMPLETED: March 2006

ANNUAL PURCHASED ENERGY USE (BASED ON SIMULATION): 202 kBtu/ft² (2,300 MJ/m²)

ANNUAL CARBON FOOTPRINT (PREDICTED): 43 lbs. CO_2/ft² (211 kg CO_2/m²

PROGRAM: Research laboratories and offices for permanent and visiting scientists.

TEAM

OWNER: U.S. Department of Energy

ARCHITECT: SmithGroup

LANDSCAPE: Andrea Cochran Landscape Architecture

ENGINEERS: Rutherford & Chekene (structural and civil); Gayner Engineers (mechanical and electrical)

COMMISSIONING AGENT: CH2M Hill

ACOUSTICS: Colin Gordon and Associates

LABORATORY: Earl Walls Associates

CONSTRUCTION MANAGEMENT: Rudolph and Sletten

LEED SCORES
LEED-NC Version 2 Platinum

	POINTS ACHIEVED	POSSIBLE POINTS
SITES [SS]	7	14
WATER [WE]	3	5
ENERGY [EA]	9	17
MATERIALS [MR]	6	13
INDOORS [EQ]	9	15
INNOVATION [ID]	5	5

MOLECULAR FOUNDRY
U.S. DEPARTMENT
OF ENERGY
BERKELEY, CALIFORNIA

ALEX WILSON

The glazing on the dramatic, west-facing cantilevered facade sports silkscreening to reduce unwanted heat gain beyond the shading provided by the spectrally selective, low-e coatings that are used throughout the building.

TIM HURSLEY

Structural Engineer C. Mark Saunders, president of Rutherford & Chekene Consulting Engineers, notes the site's high seismic force level, due to its proximity to the Hayward Fault, also complicated the design, as did the architect's decision to cantilever the top four floors 45 feet out over the hillside. "For reasons of seismic performance, we did not want to tie the building superstructure into the hillside," says Saunders. The site excavation was shored using soldier piles and 70-foot-long drilled tie-backs, and the shored wall was faced with shotcrete. To keep it from sliding down the hill in an earthquake, the building is anchored at its base with three-foot-diameter, cast-in-place concrete piers extending about 50 feet into the ground below the first floor.

A 12-foot-deep truss anchors the cantilevered floors from above, according to Saunders. "The truss served the dual purpose of supporting the floors below and supporting the screen for mechanical equipment located on the roof," he says.

The program for the building also posed significant challenges, according to Napier. Extremely low vibration, acoustic isolation, low electromagnetic interference, and super-clean laboratory environments (Class 1000 and Class 100 clean rooms) were among design requirements. "Vibration was the biggest issue," says Nick Mironov, principal of Gayner Engineers, which handled the mechanical engineering. Mironov kept most of the rotating equipment out of the building altogether, placing it in a utility building that was built simultaneously. Only ventilation air-handling equipment was kept in the foundry building. Burying portions of the lower two floors into the steep slope also helped to satisfy the need for vibration control and sound isolation.

Steve Greenberg, who was one of the LBNL researchers active on the design team, points to challenges created by the highly varied laboratory functions required of the building. "With 20 design firms and consultants participating in the process, in addition to LBNL facilities users representatives, the group required significant coordination," he says. Relative to greening, "keeping the goals in line with costs was an ongoing challenge," says Napier. Greenberg notes the sustainability elements and LEED certification were not in the original budget; it was necessary to get buy-in from the building users that these aspects to the project were important.

Once buy-in was achieved, an integrated team effectively addressed green features. Gayner Engineers, for example, was involved right from the programming stage. In part due to this integrated design, the project has become a leading model for green, energy-efficient laboratory buildings. It is a featured

Bamboo flooring and FSC-certified woods were used in public spaces and informal meeting areas, including the interior reception area shown above, lower image.

The sloping hillside location proffers long views and generous daylighting in many offices and labs.

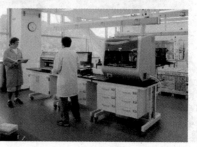

«
A brightly colored stairwell stretches from the bottom to the top of the building. It is shown here at the third level, adjacent to the main entrance.

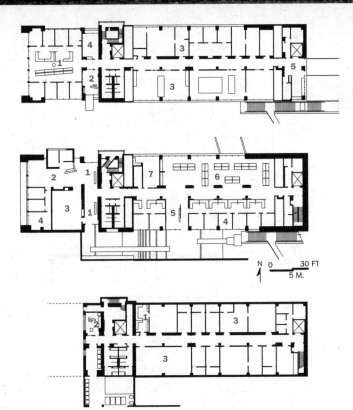

FIFTH FLOOR

1 Offices
2 Interaction
3 Laboratories
4 Conference
5 Entry, 5th floor

THIRD FLOOR

1 Main entry
2 Lobby
3 Seminar room
4 Offices
5 Interaction
6 Open office
7 Conference

N   0   30 FT
         5 M.

FIRST FLOOR

1 Offices
2 Interaction
3 Imaging laboratories

case study in the pilot Laboratories for the 21st Century (Labs21) program of the Department of Energy and Environmental Protection Agency, and has earned a LEED Platinum rating.

Dealing with construction-related indoor air quality requirements was especially challenging, according to Albert Lee, the construction manager for general contractor Rudolph and Sletten. For example, blowing out the ductwork was a challenge because the air handlers for the building divide the space vertically, serving all floors, while the building was constructed floor-by-floor.

Air flow is typically the number one energy consumer in a laboratory building. The fume hoods are variable-volume and use combination horizontal and vertical sashes, according to Greenberg, to ensure safety to researchers while minimizing the amount of ventilation air that needs to be heated or cooled. Common air-handling units for the office and lab areas provide nearly 100-percent outside air to offices, with the office air cascaded into the labs. Nearly all fans and pumps in the facility use variable-frequency drives and are controlled with reset schedules to minimize energy use for air and water flows. Energy was also saved, according to Mironov, by specifying electrostatic filtration instead of conventional bag filtration, thereby reducing static pressure drop and blower energy consumption.

The "right-sizing" of mechanical and electrical systems resulted in 30 to 40 percent reductions in equipment sizes for air handling, water heating, cooling, and electrical supply, generating first-cost savings of about $4 million, according to Greenberg, which "more than paid for the extra cost of the other greening features and the LEED process." The building is predicted to perform 28 percent better than California Title 24 requirements, which is good for half of the LEED energy credits, says Greenberg.

Other green features of the foundry include extensive access to daylighting; bamboo flooring and

>>
Among the various labs are some supporting photosensitive processes, such as the nanolithography lab (right, top), which is illuminated with orange-yellow light. The interiors were designed to provide views into many of the labs to promote communication and collaboration (right, bottom). The south-facing outdoor patio (far right) is on the third-floor level, halfway up the slope that provides first floor access at the western end and fifth-floor access at the east.

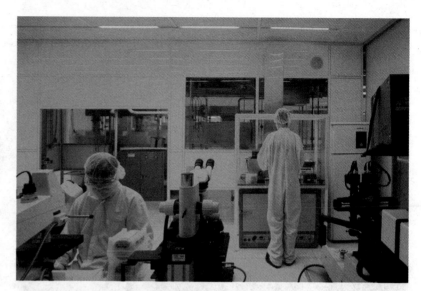

SITE PLAN

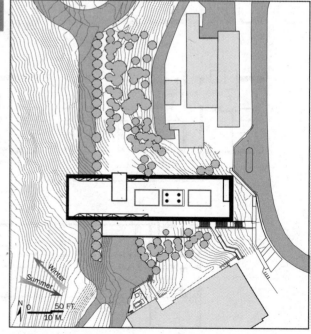

TIM HURSLEY

SKY CONDITIONS

Skies in Berkeley are relatively clear year-round, creating a direct solar load that must be carefully managed.

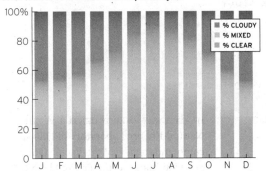

TEMPERATURES & DEW POINTS

Daily and seasonal temperature variations are quite small.

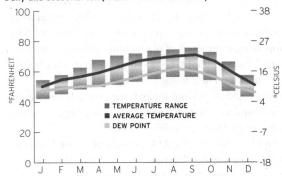

HEATING/COOLING DEGREE DAYS

Moderate temperatures provide comfortable outdoor conditions much of the year, but internal gains from laboratory equipment generates a sizable cooling load.

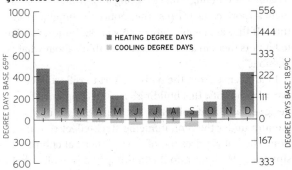

cabinetry in interaction spaces; Forest Stewardship Certified (FSC)-certified wood throughout; low-VOC-emitting carpet, paint, sealant, and adhesives; 0.5-gallon-per-minute lavatory faucets; waterless urinals; electromagnetic water treatment for the cooling towers to reduce water consumption and the use of harmful chemicals; the recycling of over 80 percent of construction waste; availability of bicycle racks; and landscaping with native plants.

The designers considered various flooring and countertop options, including natural linoleum. "The material that best stood up to their intensive chemical use, including nitrogen, was sheet vinyl—not a very green product," says Napier. One factor with flooring was the H-8 laboratory occupancy standard (specific to California) that requires spill containment with a floor covering that lines the walls. Polished concrete flooring was used in some public spaces, including the primary entry, but porosity and risk of biocontamination precluded the use of concrete floors in laboratory areas. For countertops, they used phenolic resin material instead of epoxy everywhere except in fume hoods where the harshest chemicals will be used.

As of early 2008 the building is 90 percent occupied. Starting in early 2007, LBNL began measuring energy performance—not only to help gauge the performance of this building, but also to help in setting goals for planning its next building. «

DAVID WAKELEY

SOURCES

METAL/GLASS CURTAINWALL:
Elward Metal Panel System;
Kawneer Curtain Wall
Systems

GLASS: Viracon

DOORS: Marshfield Wood

CABINETWORK AND CUSTOM
WOODWORK: Isec Corporation
Custom Casework

PAINTS AND STAINS:
Pittsburgh Paints

PANELING: Smith & Fong
Plyboo Bamboo Paneling

FLOORING: Smith & Fong
Plyboo Bamboo Flooring;
Nora Rubber

CARPET: Shaw Contract

ELEVATORS: Otis Gen2

HAWAII GATEWAY ENERGY CENTER AT NELHA, KAILUA-KONA, HAWAII

TRISTAN KORTHALS ALTES

HE GATEWAY ENERGY CENTER AT THE Natural Energy Laboratory of Hawaii commands a view of the Pacific Ocean any Hawaiian vacationer would envy. The 3,600-square-foot center consists of two buildings side by side: a conference and educational center and a smaller administrative space. But what overwhelms the building, appearing like a technological interloper in a natural ecosystem, is a white steel truss system supporting several sets of photovoltaic panels—some of them pointing up into the sky from the roof of the building, some of them shading the project's front patio entrance and steps from the parking area. Altogether, the 20 kW grid-tied system is currently providing 10 percent more energy than the building needs.

Architect Bill Brooks, AIA, of Honolulu-based Ferraro Choi and Associates, intended the design of Hawaii Gateway Energy Center in Kailua-Kona, Hawaii, to catch the eye of passersby. Will Rolston, the Hawaii Gateway manager, reports it's working. "Visitors come up the stairs with their eyes wide open," says Rolston, noting that, according to a survey, 70 percent of visitors, many of them passing the building while traveling to or from the island's main airport, come because they notice the unique structure. "If they are people with energy backgrounds, you can tell it puts them in a place where they think about what is possible."

In a reversal from the typical effort to reflect heat away from roofs, this building's curved copper roof is designed as a heat collector. The sun heats air in a plenum under the roof, inducing stack-effect ventilation. The hot air rises out of a set of thermal chimneys, siphoning fresh air into the building at a rate of 12 to 15 air changes per hour through an underfloor plenum. That air enters the building through a small exterior structure that houses coils containing 45°F seawater pumped from 3,000 feet below the ocean's surface. The

LEED SCORES
LEED-NC Version 2 Platinum

	Achieved	Possible
SITES [SS]	9	14
WATER [WE]	5	5
ENERGY [EA]	17	17
MATERIALS [MR]	4	13
INDOORS [EQ]	12	15
INNOVATION [ID]	5	5

■ POINTS ACHIEVED   ▨ POSSIBLE POINTS

KEY PARAMETERS:

LOCATION: Big Island of Hawaii, (western shore)

GROSS SQUARE FOOTAGE: 3,600 ft² (335 m²)

COST: $3.5 million

COMPLETED: October 2004

ANNUAL PURCHASED ENERGY USE (EXTRAPOLATED FROM FIVE MONTHS ACTUAL USAGE): 3.5 kBtu/ft² (-39 MJ/m²)

ANNUAL CARBON FOOTPRINT (PREDICTED): 2 lbs. CO_2/ft² (-9 kg CO_2/m²)

PROGRAM: Offices, conference facilities, labs, visitors' center

TEAM

OWNER: Natural Energy Laboratory of Hawaii Authority

ARCHITECT/INTERIOR DESIGNER: Ferraro Choi and Associates

LANDSCAPE: LP&D Hawaii

ENGINEERS: Libbey Heywood (structural); Lincolne Scott (MEP); R.M. Towill (civil)

COMMISSIONING AGENT: Engineering Economics

ENVIRONMENTAL/ ENERGY CONSULTANT: Lincolne Scott

SPACE FRAMES: Triodetic Space Frames

LEED MANAGEMENT: RMI/ENSAR

PHOTOVOLTAICS: Hawaii Electric Light Company

GENERAL CONTRACTOR: Bolton

FRANZEN PHOTOGRAPHY

Thermal chimneys emerging from the building's roofs help power a passive cooling system that reduces energy needs, while a photovoltaic array supplies 110 percent of the energy needs.

Gateway to Sustainability

THIS LEED-PLATINUM CENTER TO PROMOTE RENEWABLE ENERGY IN HAWAII
RUNS ON THE "EARTH'S DEVICES"

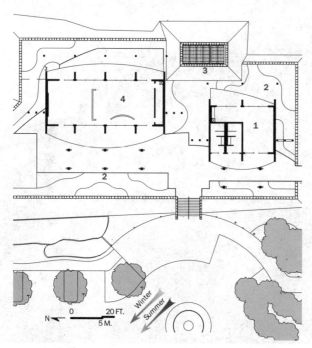

The center supports education and networking for Hawaii's renewable energy future. Registers at the perimeter of rooms deliver air cooled by thermal energy from seawater.

SITE PLAN

1 Administration
2 Landscape with condensation irrigation
3 Fresh-air inlet and deep sea cooling coils
4 Multipurpose room

4

3

2

1

2

N   0   20 FT.
        5 M.

Winter
Summer

SOURCES

GLASS: PPG, Azurlite Aqua-Blue

CEILINGS: Tectum acoustic ceiling planks

FLOORING: Tajima Free-Lay Vinyl tile with 100 percent post-consumer recycled content

CARPET: Interface tile with 50 percent recycled content, Paesaggio and Paintbox series

SPECIAL SURFACING: Yemm & Hart, Origins, polyethylene sheet of 100 percent post-consumer recycled resin

Surrounded by natural landscaping and Hawaii's *pahoehoe* lava, the center's striking appearance brings in visitors who want to learn more.

coils cool and dehumidify the air. Water that condenses on the coils drips into a collection system and is used to flush toilets and irrigate plants. Despite Hawaii's hot, humid climate, "the building is pleasant, almost too cool for some people," says Rolston.

The innovative passive cooling system changed radically over the design process, says engineer Shayne Rolfe of Lincolne Scott, the project's mechanical, electrical, and plumbing consultants. "We first put together a concept sketch with some preliminary modeling. At that point it was really a square building with a 64-foot-high thermal chimney sticking out the top," Rolfe says. "Bill came back with the idea of laying it down a little, so that we had an angular chimney." The engineer embraced the design of multiple chimneys, each three feet in diameter, which, with the angularity, helped improve the system's functionality. "From then on, it became more of a conventional mechanical design process," Rolf says.

Although the truss system appears ready-made for the photovoltaic panels it holds, it was originally designed to support long chimneys. When the plenum under the roof became part of the thermal chimney, computer simulations showed the chimneys did not need to protrude far enough to warrant the trusses. Meanwhile, the client had arranged with the local utility to provide the photovoltaic array and the trusses again had a use.

The passive conditioning system was made possible by the availability of cold seawater, offered by the state to the renewable energy campus and adjoining aquaculture facilities for $0.32 per thousand gallons. A pump circulates the seawater as needed through the cooling coils, representing the only moving part and only electricity use in the space-conditioning system.

Experience with the facility has shown that in a south wind, which is most common, the thermal chimneys work well. The air exchange rate is too low in a north wind, however. The photovoltaic panels, not originally present in computer modeling, deflect the north wind into the chimneys, counteracting their draw. The designers are working on modifications to resolve this problem. The airflow rate also decreases under cloudy skies, but so does the need for cooling. "You can really notice the building breathing when the sun goes behind clouds," says Rolfe.

To reduce solar penetration in this hot climate, the project team oriented the energy center on an east-west axis. Daylighting provides all of the building's lighting needs during business hours, with overhangs blocking direct sunlight. Occupancy and daylight sensors control the lights, which never come on during daytime, according to Rolston. Located on the barren landscape of Hawaii's *pahoehoe* lava, the building does not disturb the site beyond its footprint. For irrigation of native plantings, the project again uses the thermal energy

SKY CONDITIONS

Cloudiness data was not available for Kailua-Kona, so this chart represents conditions in Honolulu.

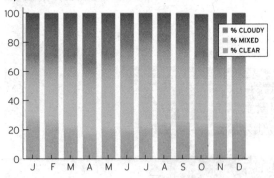

Legend:
- % CLOUDY
- % MIXED
- % CLEAR

TEMPERATURES & DEW POINTS

Hawaii is characterized by consistently warm temperatures year-round.

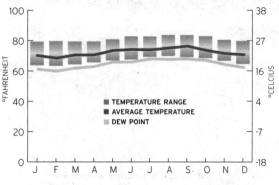

Legend:
- TEMPERATURE RANGE
- AVERAGE TEMPERATURE
- DEW POINT

HEATING/COOLING DEGREE DAYS

Heating is never required at this site, so the only comfort load is for cooling.

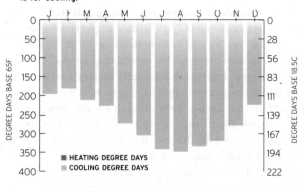

Legend:
- HEATING DEGREE DAYS
- COOLING DEGREE DAYS

A project with this many innovations could only result from a fully integrated design process.

▲ Under the hot sun, the center's passive cooling system delivers 12 to 15 air changes per hour, while dehumidifying the air. That rate is reduced during cooler weather and at night, when cooling needs are less.

of the cold seawater. The plants are watered with vapor that has condensed on cold-seawater pipes running over the ground like a drip irrigation system but with water dripping off of the pipes instead of out of them.

A project with this many innovations could only result from a fully integrated design process. The client wanted an environmentally responsible building, with a request for proposals that emphasized sustainability. "We had a team that was purposefully constructed to be highly motivated to do something sustainable," says Brooks. Even without experience on green projects, a contractor based on the island beat several off-island contractors for the job. "They became totally invested in the idea that they were one of the first contractors in the state working on a sustainable project," says Brooks. About the engineering firm, Lincolne Scott, Brooks adds, "They came from a mind-set of 'Let's go for it; let's do something different.' "

Completing the project within its $3 million budget was a challenge, in part due to construction costs. "It's a contractor's market on the island," says Brooks, noting the limited pool of contractors and the need to import building products. The team found most LEED credits were applicable to the project, making LEED Platinum achievable, but the building performed poorly in the materials-and-resources category. "To use local materials [which can earn LEED credits], you're restricted to a few selections, which normally boil down to concrete," says Brooks. "Everything else is imported."

Despite the building's exemplary energy performance, the project faced an obstacle attaining LEED Platinum in energy use. "LEED calculation doesn't account for passive ventilation systems that work like mechanical systems,"

158 EMERALD ARCHITECTURE: CASE STUDIES IN GREEN BUILDING

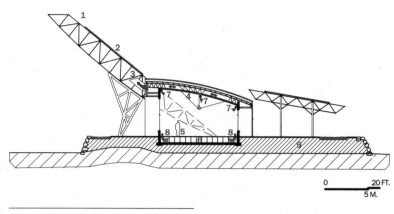

TYPICAL CROSS SECTION

1 North space frame
2 Photovoltiacs array
3 Hot-air exhaust outlet into "chimney"
4 Insulated ceiling plenum
5 Access flooring

6 Cool, fresh air enters
7 Exhaust inlet
8 Cool air outlet
9 Hawaiian *heiau*
  (traditional platform)

SEAWATER PUMP DESIGN

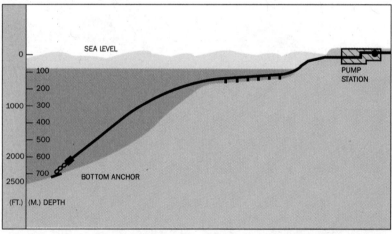

says Brooks. Since the building has no mechanical cooling system, LEED compared it with buildings with no cooling systems at all, which negated the energy savings achieved by its passive system. The project achieved the maximum number of energy points possible by leaning on its renewable energy generation, Brooks notes. According to the energy analysis performed by Lincolne Scott, the building performs 54 percent better than an ASHRAE 1999-90.1 base case using a conventional cooling system.

The building was commissioned by Environmental Economics, although, says Brooks, "there was not that much to commission since our building is designed to move air without moving parts." The commissioning called for some corrections with daylight sensors and lighting, and the process also resulted in refinements to the rate of flow and pressure in the on-site seawater pump, resulting in significant energy savings. "The whole idea was to circulate only enough seawater to allow proper cooling," says Brooks, and the team is still refining that balance. "We think [pumping energy] can still go down 10 percent to 15 percent," he notes.

The center won an AIA/COTE Top 10 award for 2007. Speaking for the jury, Traci Rose Rider said, "We were impressed by the way they blended active and passive technologies. It's really using all of earth's devices, then dramatizing that with this visible structure."

With the center fully booked with educational sessions, the attention garnered by its architecture has also brought new awareness to the need for renewable power in Hawaii. Even passing aircraft seem to change course to get a better look at the building, says Rolston, noting, "Sitting out there on the lava, it's an interesting thing to see." «

Especially visible at dusk, the solar array contributes drama to the stark landscape.

Index